POWERING
the
FUTURE

How We Will (Eventually)
Solve the Energy Crisis
and Fuel the Civilization
of Tomorrow

Robert B. Laughlin

NOBEL LAUREATE IN PHYSICS

BASIC BOOKS
A MEMBER OF THE PERSEUS BOOKS GROUP
New York

Copyright © 2011 by Robert B. Laughlin
Published by Basic Books,
A Member of the Perseus Books Group

Books published by Basic Books are available at special discounts for bulk
purchases in the United States by corporations, institutions, and other
organizations. For more information, please contact the Special Markets
Department at the Perseus Books Group, 2300 Chestnut Street, Suite 200,
Philadelphia, PA 19103, or call (800) 810-4145, ext. 5000, or e-mail
special.markets@perseusbooks.com.

Designed by Timm Bryson

The Library of Congress has cataloged the hardcover as follows:
Laughlin, Robert B.
  Powering the future : how we will (eventually) solve the energy crisis and fuel
the civilization of tomorrow / Robert B. Laughlin.
    p. cm.
  Includes bibliographical references and index.
  ISBN 978-0-465-02219-9 (hardback) — ISBN 978-0-465-02794-1 (e-book)
1. Renewable energy sources—Popular works. 2. Power resources—Popular
works. I. Title.
  TJ808.L365 2011
  333.79—dc23
                                    2011024618

10 9 8 7 6 5 4 3 2 1

To my sons—and others of future generations
whose problem this will be

CONTENTS

CHAPTER 1
Armchair Journey

A few hundred years from now, human beings won't burn coal, oil, and natural gas any more. This might be because they've adopted laws that keep supplies in the ground for environmental reasons, or it might be because supplies are all used up, but this distinction is a detail. The amount of fossil fuel in the earth is finite, so people eventually won't be able to burn it any more.[1] We don't have to analyze contemporary energy struggles and pilot our way through the many minefields and crossfires to see the energy future in this respect. We can skip over all that and simply transport ourselves mentally to a time when fossil fuels are gone.

Let us therefore take an armchair journey into the distant future, when coal, oil, and natural gas are things of the past. There's a slight possibility that the trip will fail because the entire human race got wiped out in an earlier environmental catastrophe or war, so we find no people, but that's extremely unlikely. Humans are very adaptive and prolific creatures, and every last one of us would have to be eliminated to prevent us from repopulating. So let's assume this didn't happen, and that we encounter some people not so different from ourselves, for they're carbon copies of us, literally.

Partisans in modern energy wars are quick to point out how irresponsible such thinking is. The final demise of carbon burning is so far away, perhaps ten generations, that it's quite irrelevant to energy problems of today.[2] They say that by focusing on the distant future, we only encourage apathy.

However, transporting ourselves mentally beyond any living person's self-interest has the great advantage of separating technical matters from political

1

ones. Modern energy concerns are inherently political, of course, so divorcing the two completely makes no sense. But we save ourselves much time and vexation if we deal with the much simpler technical issues first. Think of it this way: To build a power plant, we need both enough votes and enough concrete, but if there isn't any concrete, we're simply not going to build the plant.

Moreover there's more benefit than usual in seeking clarity in this case because the very large stakes involved encourage people to misunderstand each other and, dare we say it, make a statement or two now and then that they know to be untrue. For instance, *we don't need alternative fuels because the world isn't running out of oil, oil is plentiful right now, we're making new discoveries all the time, and there are forty-two years' worth left.*[3] Or, *we do need alternative fuels because the world is running out of oil, it will happen in a decade or two, and we need to invest in green technologies and distance ourselves from coal.* Or, *we need fast trains because the world is running out of oil, big cars are wasteful, the trains will go everywhere we want, and taxing truckers is good.* All sides in such arguments tend to represent that science is on their side and that theirs is the only logically sound position. Logic has nothing to do with it, of course.

The issues of energy turn out to be especially easy to think through in hindsight. For example, we might ask whether the people we encounter in this distant time will still drive cars. After a few moments of thought, nearly everybody answers yes to this question, even though they're not quite sure where the energy would come from. The reason, they say, is that cars will be things that people desperately want, if only because they're status symbols, and they'll therefore pay any price, whereupon entrepreneurs will step forward to find a way to make the cars available. We can also ask whether these people of the distant future will still fly in airplanes. That's a little harder, for it's easier for us to imagine living without airplanes than without cars. However, again nearly everyone concludes that people of means living at that time will want the speed and convenience of air travel, and thus that ordinary folk will be able to fly too, although not necessarily cheaply. Then there's the question of whether the lights will come on—that is, whether electricity will be available at reasonable prices whenever users want it. Everyone answers this one yes very quickly, reasoning that governments foolish enough to let the lights go out will not be in office for long.

With the basic features of the future energy landscape thus determined, important technical details now fill in easily. If people are flying in airplanes, they must have something to power those planes. It can't be petroleum distillates because there isn't any petroleum. The fuel, whatever it is, has to be light, compact, and safe because otherwise the airplanes won't fly—or will blow up occasionally if they do fly. The only such substance elementary chemistry allows is the very jet fuel we use today, so these people must be synthesizing jet fuel from raw materials, presumably with help from an outside energy supply of some kind. Likewise, if people are driving cars, they must be powering them somehow. The power source might also be synthetic fuel, or it might be something else, such as batteries or third rails, but it will definitely be the least expensive option. These people won't like wasting money any more than we do, and they especially won't like wasting it on energy companies. If people's lights come on when they flip the switch, then the power again must come from somewhere. It could be from the sun, the wind, or nuclear reactors, but where it will actually come from is the producer with the lowest delivery price.

We might worry at this point that these people fibbed to us about their world having sufficient energy resources to cover their needs, but a quick check dispels that worry. There is more than enough supply by a wide margin, notably from the sun and its proxy, the wind. It's just a matter of what has the lowest cost. Not only have these folks not fibbed, but in fact they seem to be stuck in exactly the same rat race of production and delivery cost minimization that we are in today, only with different details.

The nature of the future energy enterprise our armchair journey reveals isn't directly relevant to any present-day energy controversy, for we live in a time when fossil fuels dominate prices. It matters peripherally, however, because the seeds of what we should do now are contained in what will be. If, for example, we think that carbon is destined to play a central role in energy long after fossil fuels are gone because it's indispensable for air travel, we might want to start being nicer to our carbon industries. If we think that synthetic fuel manufacture is destined to materialize no matter what, we might want to encourage its creation now, so that we don't have to build plants in a panic when the crisis is upon us. If we think that nuclear power is destined to impose a price ceiling on electricity no matter what, we might want to develop it properly and keep it standing by, whether we deploy it or not. If we think solar and wind power are destined to be a central energy source no matter what, we might want to develop ways to bank the energy they produce in extremely large quantities, even though doing so is expensive.

We can predict the energy situation so far in the future reasonably reliably because it's circumscribed by elementary things. In this it's different from, say, the weather or an election return. We know that the laws of economics will still hold, even if the direst predictions of global warming come true, and even if there is serious military conflict between now and then. The people of that time will be just as selfish as we are, just as ambitious, just as motivated to protect the children, and just as clever, thanks to the magic of genetics. Also, energy differs from other aspects of human economic life, such as market presence or stock value, in being extremely primal and ruled by simple, powerful physical law. We know that this law has been in force since long before humans walked the earth, and we're reasonably confident that it will never change or be overturned by future discoveries or technical innovations. The equations of quantum mechanics will be exactly the same centuries from now as they are today, regardless of what happens, as will the rules of chemistry and engineering that flow from them, as will the equations of heat, light, electricity, and radiation. Energy will still be conserved and will still have to come from somewhere. Its possible sources will be the same as they are today in every detail. It will still degrade entropically with use and have to be thrown away into outer space as waste heat when its work is done. People of that day will still need a steady supply of it to live.

However, the simplicity of this energy calculation is a double-edged sword, for it obligates us to check our facts carefully. We need to verify that present-day energy economics really is as cutthroat as everyone says it is. We need to verify that quantum mechanics really does prevent stressed matter from ever matching chemical fuels as energy storage media. We need to check that ordinary jet fuel really does have the optimal energy density allowed by physical law and that the carbon it contains is essential. We need to verify that electricity and magnetism are unsurpassed for transmitting energy but useless for storing it. We need to scrutinize the costs of batteries, including the hidden costs associated with the environmentally unfriendly metal ions they contain. We need to recheck our nuclear facts, particularly as regards supply, costs, and the burdens of radioactive waste disposal and antiterrorism security. We need to scrutinize biological processing costs and make sure they're compared properly with those of conventional chemical processing. And finally, we need to review and pin down, as best we can, the supply and cost numbers of solar energy in its various forms.

Thinking about the end of fossil fuels futuristically also makes painfully clear that climate and energy are very different things. The people working in the twilight of the fossil fuel era will care about the earth as much as we do, but they'll be struggling to live by tasking earth's resources in new ways and will have to make that struggle their first priority. Their cost constraints will also be especially severe because they won't have cheap fossil fuel to fall back on anymore. They'll have to pay out of pocket for expensive decisions favoring the environment. The effect that has on behavior is unhappily familiar. Individuals passionate about improving the state of the world centuries in the future usually lose interest the moment one begins enumerating specific sacrifices they should make to accomplish the improvement, especially if those sacrifices involve disadvantaging their own children. It turns out that they were only interested in saving the world with someone else's money. Such attitudes, common now, will only become more common as fossil fuel supplies wane and competition for the necessities of life intensifies. We can thus expect the move to secure adequate energy supplies, when it eventually occurs, to be largely divorced from environmental concerns or even at odds with them, because people will be putting their own needs first. Alternate energy, when it finally arrives, will not be green.

Climate and the environment will thus continue to be problematic through the transition away from fossil fuels, notwithstanding the cessation of carbon dioxide buildup in the air and the plateauing of global temperature increases. The core problem, ever-increasing human demands on the earth's resources, will still be present and may even have worsened. People of this future time will still want clean air and water, natural surroundings, large tracts of land kept wild, fisheries and forests kept healthy, and factories somewhere other than their back yards. They might not be able to have all these things, however, perhaps because the earth is warmer, but mostly because these things will come at a cost in economic well-being (for some) and because they'll be incompatible with large populations. Human burdens on the earth are a specific case of burdens that plants and animals impose generally on their respective ecological niches. The earth's limited ability to sustain such burdens is a major factor in the fight for survival and thus the rise and fall as species over the span of geologic time.

Even though the final outcome of the transition away from fossil fuels is likely to be positive, the transition itself could be terrible. It will be a major global event, like an ice age or a comet impact, not a mere budgeting shortfall or overcrowded highway that an act of Congress can whisk away.[4] The present flow of crude oil out of the ground, the greatest since the oil age began, equals in one day the average flow of the Mississippi River past New Orleans in thirteen minutes.[5] If we add the energy equivalents of gas and coal to the calculation, it's thirty-six minutes. Although daring use of capital was essential for harvesting this abundance, it didn't put the abundance there in the first place and won't make the abundance reappear once it runs out. When it does, people won't merely be inconvenienced. Rather, the earth's capacity to render up unimaginably large amounts of oil, gas, and coal on demand is a fundamental premise of modern industrial civilization. Without exception, all major cities in the world are too large to feed without the use of machines. Were energy supplies to fail catastrophically tomorrow, the great city as we know it would cease to exist, and most of us would starve. The decline and eventual exhaustion of fossil fuels is thus like the coming of winter to a people who have known only summer.

What often occurs when humans compete for diminishing resources is military conflict. However, it's best not to dwell on that.[6]

CHAPTER 2

Geologic Time

Ironically, a proper mental journey into the energy future must begin with a backward look at geologic time. The reason for this is that the way forward is fogged by misunderstandings about the earth, the starting point for any serious conversation about energy supply or climate. We're constantly struggling to separate myth from reality when thinking about these things. Experts are little help, for they have the same difficulty, which they routinely demonstrate by talking past each other. Respected scientists warn of imminent energy shortages as geologic fuel supplies run out. Wall Street executives dismiss their predictions as myths and call for more drilling. Environmentalists describe the destruction to the earth from burning coal, oil, and natural gas. Economists describe the danger to the earth of failing to burn coal, oil, and natural gas. Geology researchers report fresh findings about what the earth was like millions of years ago. Creationist researchers report fresh findings that the earth didn't exist millions of years ago. The only way not to get lost in this awful swamp is to review the basics and decide for yourself what you believe and what you don't.

Geologic time is such a vast concept that converting it to something more pedestrian, just to get oriented, can be helpful for our understanding. I like rainfall. The total precipitation that falls on the world in one year is about one meter of rain, the height of a golden retriever.[7] The total amount of rain that has fallen on the world since the industrial revolution began is about two hundred meters, the height of Hoover Dam. The amount of rain that has fallen on the world since the time of Moses is enough to fill up all the oceans.

The amount of rain that has fallen on the world since the ice age ended is enough to fill up all the oceans four times. The amount of rain that has fallen on the world since the dinosaurs died is enough to fill up all the oceans twenty thousand times—or the entire volume of the earth thirty times. The amount of rain that has fallen on the world since coal formed is enough to fill up the earth one hundred and forty times. The amount of rain that has fallen on the world since oxygen formed is enough to fill the earth one thousand times.[8]

Common sense tells us that damaging a thing this old is somewhat easier to imagine than it is to accomplish—like invading Russia. The earth has suffered mass volcanic explosions, floodings, meteor impacts, mountain building, and all manner of other abuses greater than anything people could inflict, and it's still here. It's a survivor. We don't know exactly how the earth recovered from these devastations, for the rocks don't say very much about that, but we do know that it did recover, the proof being that we are here.

Nonetheless, damaging the earth is precisely what's concerning a lot of responsible people at the moment. Carbon dioxide from human fossil fuel burning is presently building up in the atmosphere at a frightening pace, enough to double the present concentration in a century. Some people predict that this buildup will raise average temperatures on earth several degrees centigrade, enough to modify the weather and accelerate the melting of the polar ice sheets.[9] Governments around the world have become so alarmed at this prospect that they've taken significant (although ineffective) steps to slow the warming. These actions include legislating carbon caps, funding carbon sequestration research, subsidizing alternate energy technologies, and initiating at least one serious international treaty process to balance the necessary economic sacrifices across borders.[10]

Unfortunately, this concern isn't reciprocated. On the scales of time relevant to itself, the earth doesn't care about any of these governments or their legislation. It doesn't care whether we turn off our air conditioners, refrigerators, and television sets. It doesn't notice when we turn down our thermostats and drive hybrid cars. These actions simply spread the pain over a few centuries, the bat of an eyelash as far as the earth is concerned, and leave the end result exactly the same: All the fossil fuel that used to be in the ground is now in the air, and none is left to burn.[11] The earth plans to dissolve the bulk of this carbon dioxide into its oceans in about a millennium, leaving the concentration in the atmosphere slightly higher than today's. Over tens of millennia after that, or perhaps hundreds, it will then slowly transfer the excess carbon dioxide into its rocks, eventually returning levels in the sea and air to what they were before humans arrived on the scene. The process will take an eternity from the human perspective, but it will nonetheless be only a brief instant of geologic time.

Some details of this particular carbon dioxide scenario are controversial, of course, because all forecasts are partly subjective, including those made by computer. We have to extrapolate from present-day facts and principles, and there are varying opinions about these. The time scale for the ocean to absorb man-made carbon dioxide is set by the mixing rate of surface water with deep water in the sea, which is known only indirectly and might conceivably change during the thousand-year heating transient.[12] The amount of carbon dioxide left in the atmosphere after equilibration varies from tolerable to alarming depending on how much industrial burning you assume in your model.[13] No one knows for sure how long it will take the excess carbon dioxide to disappear into the rocks or even the specific chemistry involved. The main reason for thinking it will disappear is that something, presumably a geologic regulatory process, fixed the world's carbon dioxide levels before humans arrived on the scene. Some people even argue that carbon dioxide has been fixed at these values for millions of years, the grounds being that the photosynthetic machinery of plants seems optimized to them.[14] But the overall picture of a thousand-year carbon dioxide pulse followed by glacially slow decay back to the precivilization situation is common to most models, even very pessimistic ones.

Global warming forecasts have the further difficulty that one can't find much actual global warming in present-day weather observations. In principle, changes in climate should show up in rainfall statistics, hurricane frequencies, temperature records, and so forth. As a practical matter they don't (yet) because weather patterns are dominated by large multiyear events in the oceans, such as the El Niño Southern Oscillation and the North Pacific Gyre Oscillation, that have nothing to do with climate change.[15] In order to test the predictions we'd have to separate these big effects from subtle, inexorable changes on scales of centuries, and nobody knows how to do that at the moment.

Humans can unquestionably do damage persisting for geologic time if one counts their contribution to biodiversity loss. There is considerable evidence that humans are presently causing something biologists call the "sixth mass extinction," an allusion to the five previous cases in the rocks in which huge numbers of species died out mysteriously in a flash of geologic time.[16] A popular—and plausible—explanation for the last of these events, the one when the dinosaurs disappeared, is that an asteroid ten kilometers in diameter struck the earth traveling fifteen kilometers per second and exploded with the power of a million 100-megaton hydrogen warheads.[17] Many say that the damage that human activity presently inflicts is comparable to this. Extinctions, unlike carbon dioxide excesses, are permanent. The earth didn't replace the dinosaurs after they died, notwithstanding the improved weather conditions and twenty thousand ages of Moses to make repairs. It just moved on and became something different than it had been before.

However, carbon dioxide per se is not responsible for most of this extinction stress. There are a handful of potential counterexamples, such as corals,[18] which seem especially sensitive to acidification of the ocean surface, and amphibians, which have been declining noticeably for unknown reasons.[19] But except in these few isolated cases, keeping carbon-based fuels in the ground a while longer won't do much to mitigate the loss of biodiversity. The real problem is human population pressure generally—overharvesting, habitat destruction, pesticide abuse, species invasion, and so forth. Slowing man-made extinction in a meaningful way would require drastically reducing the world's human population.[20] For better or worse, that is unlikely to happen.

It's a mistake to suspend judgment on questions of population, climate, carbon use, and so forth just because they're sensitive. If you do, you'll become incapacitated by confusion. Earth scientists tend to be ultra-conservative when it comes to the future, presumably to avoid getting tarred as myth-mongers, and they go to extraordinary lengths to prove that the globe is warming now, the ocean is acidifying now, fossil fuel is exhausting now, and so forth, by means of measurement, even though these things are self-evident in geologic time. The unhappy result is that we get more and more data but less and less understanding—a common problem in science but an especially acute problem in climatology. In such situations it's essential to weigh facts more strongly if they are simple and use this practice to sweep away confusion whenever (and if) we can.

The sea's immense capacity to store carbon dioxide is one of the simple things with which we can reliably orient ourselves. It's a junior high school science fair project. If you leave a glass of distilled water on the counter overnight, you will find the next morning that it has become slightly acid, due to absorbing carbon dioxide from the air.[21] It hasn't absorbed much—about the amount stored in an equal volume of air—so this effect alone will not sequester much carbon. But if you now drop a piece of limestone in the water, thereby emulating the presence of carbonate rocks at the bottom of the sea, and repeat the experiment, you will find that the water has become slightly alkaline and that the amount of carbon dissolved in the water is now sixty times greater than it was before.[22] You have to tinker a bit to figure out where this carbon came from, but you eventually discover that half came from the limestone and half came from the air. It all has to do with the marvelous (and elementary) chemistry of bicarbonate salts. You also find that the alkalinity matches that of seawater, as does the carbon dioxide carrying capacity. Thus we learn that the oceans presently have dissolved in them, in the form of bicarbonate ion, forty times more carbon than the atmosphere contains, a total of thirty trillion tons, or thirty times the world's coal reserves.[23]

The experiments that constrain geologic time scales are almost as simple as this science fair project, although not quite, and they orient us just as reliably. Not everyone agrees with this assessment, of course.[24] Geologic time does contravene religious beliefs, a difficulty with this subject that is very regrettable because it doesn't contravene the religious beliefs that count. But it's probably more significant that the experiments, simple though they may be, involve obscure facts about rocks, a knowledge of physical law, and the assumption that this law was the same in the ancient past as it is now, none of which is obvious, much less interesting, to the average person. If you go to the supermarket and engage the checkout clerk in a conversation about the Paleozoic era, radioactivity, or the disappearance of the megafauna, you'll be met with a smile and then probably escorted from the building as a lunatic. However, the time scales do come from somewhere concrete that can be explained simply.

We get a long way toward understanding geologic time by just disciplining ourselves not to dismiss things around us that seem to make no sense. For instance, a local beach a short drive from my home is backed by cliffs about one hundred feet high that expose alternating layers of sandstone, mudstone, and aggregate, perhaps seven in all.[25] One can tell without having attended a single geology class that these layers were formed by the action of water, the most likely candidate being the nearby ocean, especially in light of the fossilized clams entombed in some of them. Yet there they are high and dry, integrated into the rolling hills beyond, as though they were the sliced edge of a huge layer cake. The layers are also tilted, sometimes up and sometimes down, as though giants had sat down on them in some places but not others. The tilt is large enough that some clifftop planes continue downward to the beach and disappear into the ground. The cliffs are eroding. The rocks are noticeably crumbly in places, and one can see little landslides high up on the face as well as shelves and caves at the bottom where waves wash at high tide.

Once we begin noticing oddities in the rocks, we can't help but think about their implications. Layers of rocks with fossilized clams in them can only be above water now if the land rose, the sea sank, or both. Sea level has been quite constant throughout recorded history, say five thousand years, and there are no documented cases of hundred-foot rises in the land either, except for volcanoes, so the cliffs are considerably older than recorded history.[26] The tilting tells us that the land moved, regardless of what the sea did. The material forming the layers had to come from somewhere. Erosion from the cliffs themselves is really the only possibility, for there just isn't enough mud coming down local creeks and rivers to account for the sheer mass of rock, and besides, the layers are grainy and chunky, whereas the river mud isn't. But cliffs can't be made of erosion debris from themselves. The cliffs must therefore have eroded away completely and risen up again at least once, and more likely several times, judging from the layering complexity. The erosion rate of the cliffs thus sets the minimum age of the rocks. This rate appears to be about one millimeter per year, perhaps less, for the rock here is relatively hard, which means a million years to erode a kilometer, or about ten million years to erode away the shore entirely. That's sufficiently long that we don't have to allow for the ice age—which we shouldn't do anyway because we're not supposed to know about the ice age yet. The age of the rocks therefore is about two million years, or perhaps four million, just to be safe.[27]

Such crude estimates of geologic time were the best anyone could do until the 1960s, when radiometric dating of rocks became commonplace. The relative newness of this technology accounts for some of geology's credibility problems, for geologic time itself was invented one hundred years earlier and thus had plenty of time to develop a reputation for flakiness. The newness also accounts for why organized opposition to radiometric dating has grown recently, as opposed to some previous time. There is no need to impugn the reliability of a thing that does not yet exist.

Although radiodating is technically difficult—indeed impossible without sophisticated equipment—it's straightforward conceptually.[28] The method appropriate to this situation involves placing a piece of rock about the size of a golf ball in a vacuum chamber, melting the rock, collecting all the gases driven off, and measuring the total mass of the element argon these gases contain.[29] Then we dissolve the same rock in acid, do a bit of conventional wet chemistry with the solution, and measure the total mass of the element potassium it contains. The ratio of these two masses, multiplied by a certain number, is the age of the rock. The physics underlying this procedure is that potassium, which is plentiful in nearly all rocks, is slightly radioactive and decays to argon, a chemically inert element. Argon likes to escape out of rocks when they are very hot, in particular when they are melted into volcanic lava, but otherwise is trapped. A conventional volcanic rock contains no argon right after it solidifies. The amount of argon it contains right now therefore counts the number of potassium atoms that decayed since it solidified and thus the amount of time that elapsed.

Radiometric dating has to be used cautiously, however, for it's notoriously easy to do wrong. The argon levels can be artificially high, for example, because of atmospheric contamination in air pockets and grain boundaries in the rock, or they can be artificially low because the rock got overheated sometime after it formed or because the rock recrystallized or acquired inclusions of younger rock through geologic processes underground. Sedimentary rock always gives nonsense readings because it doesn't get hot when it forms and because weathering, aggregation, and metamorphism cause crystal structure changes, which corrupt the argon record.

Fortunately, the cliffs on my beach possess a layer of volcanic ash fairly high up with which they can be dated. The team that last surveyed the site chose not to date the ash directly, presumably because they didn't trust the argon levels, but instead identified it chemically with ash deposited hundreds of miles away and overlain by a layer of volcanic basalt. The basalt yielded a clean argon age of two and a half million years. Basaltic rocks higher up in the mountains behind this beach, which are older, yield an age of twenty million years. The rocks on the beach are thus somewhere between two and twenty million years old. Cross-correlation of the fossils they contain then narrows this down to about six million years, give or take a million. Thus there were no human beings on earth when the lowermost of these layers first sedimented out of the sea. Between then and now enough rain fell on the earth to fill up the oceans two thousand times.

It would be very surprising if rocks conveniently near my home had especially large geologic ages, and naturally this isn't the case. When we go through the same kinds of analysis with rocks in other parts of the world, we typically get ages that are ten to one hundred times greater than these. A particularly famous example is in the first edition of *Origin of Species*, where Charles Darwin used erosion arguments to estimate the age of the Weald, a region southeast of London curiously deficient in chalk.[30] He came up with three hundred million years. It was impossible to refine this estimate radiometrically at the time, so it's probably not surprising that he reduced his estimate by half in the second edition and eliminated all mention of the subject in the third. But his reasoning was conceptually right, and the estimate itself was close to correct. The Weald is about one hundred twenty million years old, give or take ten million.[31] It's an interesting part of England—the place where the Battle of Hastings was fought, cricket was invented, and dinosaur fossils were first discovered.[32]

The Weald is only the beginning, however, because Great Britain is extremely old. By a stroke of fortune, the entire country is a complete stack of the world's sedimentary layers tipped gently downward to the northwest and then planed level at the top.[33] The plentiful fossils in the ground, which are different in different layers, thus form narrow tracks that run roughly parallel to the coast of France. When people first discovered these tracks, they had no way to date the rocks in question, so they simply assigned names. The easternmost track became Cretaceous, after the Greek word *creta* for chalk. The next one became Jurassic, after the Jura mountains in Switzerland. The next one became Triassic, after a characteristic three-level sedimentation pattern (the Tria) found commonly in Germany. The next one became Permian, after the region of Perm in Russia. And so on and so forth. But the subsequent invention of radiodating later enabled actual ages to be assigned to these names, albeit with the precision difficulties encountered on my beach.[34] The white cliffs of Dover are seventy million years old. The clay under Oxford is one hundred fifty million years old. The rocks under Stratford-upon-Avon are two hundred million years old. The coal under Stoke-on-Trent is three hundred million years old. The Lake District is four hundred million years old. The Isle of Man is five hundred million years old. The highlands of Scotland are six hundred million years old—and more.

The oldest rocks in the world are not in Great Britain but rather in places exposed to extremes of ice-age glaciation, such as Greenland, northern Canada, and northern Finland. Here the glaciers ground off all the upper sedimentary layers to expose the primordial rocks below. Radiometric ages of these rocks begin where the geological record in Britain ends and run back an additional four billion years.[35] The oldest ages coincide with those of meteorites and moon rocks, implying that they date the birth of the earth.[36] The age of the earth isn't important for energy discussions except in establishing that cosmic events, not artistic value judgments, set the overall scale of geologic time.

The continents have moved up and down over the course of geologic time a greater distance than the sea is deep. We know this because the total thickness of sedimentary rock in some places exceeds four kilometers. Charles Darwin also observed, after dating the Weald, that the total thickness of all the sedimentary strata in England would total twenty-two kilometers if piled on top of one another. It wasn't clear at the time how literally one should interpret this fact, for nobody had mined straight down through all the layers, nor did anyone know for sure how deep the ocean was. But now the oceans have been thoroughly surveyed, and oil technologies such as echo stratigraphy and deep drilling routinely find sedimentary rock layers ten to fifteen kilometers thick.[37] The most sensational example of such thicknesses is, of course, the Grand Canyon, which required a three-kilometer uplift from sea level to be cut by the Colorado River, and which forms, together with Utah's Escalante Staircase, a total sedimentary mass ten kilometers thick.[38] The Grand Canyon also demonstrates that uplift and subsidence alternated, because it contains plant fossil layers sandwiched between marine fossil layers. Less famous but no less relevant to the vastness of geologic time is the nearby Animas River canyon, which cuts through sedimentary rock five kilometers thick. Around the world, sedimentary deposits over one kilometer thick are commonplace.[39]

Sea level has not, however, moved up and down over the course of geologic time an amount greater than the mountains are tall. We know this because marine sediments have accumulated continuously for the last six hundred million years, which they would not have done if continental erosion had stopped or the seabed had emptied. Moreover, we can work backward from clues left in the rocks to reckon what the sea level was in the geologic past.[40] This process has methodological uncertainties, for it involves judgments about how layer sequences in different parts of the world line up, what constitutes evidence for shorelines, and how the earth's crust yielded and rebounded as masses of rock came and went.[41] However, the process is accurate enough to tell us that the amount of water on the earth hasn't changed significantly over geologic time, and that the rise and fall of the oceans is adequately accounted for by the waxing and waning of the polar ice sheets and slow changes in ocean basin volumes. The sea level has had a complex and interesting history, but it has never deviated more than two hundred meters from its present value.

The sea has risen and fallen particularly vigorously over the past million years as a result of ice-age glaciation. We know this because oxygen isotope ratios in the ocean sediments vary violently with depth.[42] These ratios indirectly measure the amount of water locked up in glacial ice sheets at the time of sedimentation.[43] The sediments record nine major glacial episodes, each of which lowered the sea level by more than fifty meters and then returned it abruptly to its present value.[44] At least four of these episodes lowered the sea by more than one hundred meters. This includes the most recent one, which lowered it one hundred twenty. The amount of lowering is corroborated by uplifted coral reefs, which show growth in places that would otherwise have been impossible, on account of requiring specific water depths.[45] This lowering is also consistent with estimates of the ice mass required in order to leave behind such industrial-strength mischief as Long Island, Nantucket, and the Great Lakes—about thirty million cubic kilometers in all, or thirty million billion tons.[46]

The major glacial episodes are spectacular examples of the natural climate change that has occurred in geologic time. They took place at regular intervals of one hundred thousand years and always followed the same strange pattern of slow, steady cooling followed by abrupt warming back to conditions similar to today's. We know this because chemical records in polar ice, the patterns of which match those of the sediments, contain a signal that tracks the earth's precessional wobble.[47] The precession is a clock-like astronomical quantity, so its appearance in the ice data enables us to date the ice precisely. That, in turn, enables us to date the sediments precisely. The last glacial melting, cross-dated at fifteen thousand years ago by the radiocarbon age of wood debris the glaciers left buried as they retreated, occurred rapidly.[48] The sea rose more than one centimeter per year for ten thousand years, then stopped. The extra heat required for this melting was ten times the present energy consumption of civilization.[49] The total meltwater flow was two Amazons, or half the discharge of all the rivers in all the world.

The great ice episodes were not the only cases of natural climate change, however.[50] Six million years ago the Mediterranean Sea dried up.[51] Ninety million years ago alligators and turtles cavorted in the Arctic.[52] One hundred fifty million years ago the oceans flooded the middle of North America and preserved dinosaur bones. Three hundred million years ago northern Europe burned to a desert, and coal formed in Antarctica.[53] The great ice episodes themselves were preceded by approximately thirty smaller ones between one and two million years ago, and perhaps twice that many before that.

Nobody knows why these dramatic climate changes occurred in the ancient past. Ideas that commonly surface include perturbations to earth's orbit by other planets,[54] disruptions of ocean currents,[55] the rise and fall of greenhouse gases,[56] heat reflection by snow,[57] continental drift,[58] comet impacts, Genesis floods,[59] volcanoes, and slow changes in the irradiance of the sun.[60] None of these things has yet pointed to a scientifically sound explanation.[61] However, one thing we know for sure is that people weren't involved. There weren't enough people around during the ice episodes to matter, and there weren't people before them at all.

The geologic record as we know it thus suggests that climate is a profoundly grander thing than energy. Energy procurement is a matter of engineering and keeping the lights on under circumstances that are likely to get more difficult as time progresses. Climate change, by contrast, is a matter of geologic time, something that the earth routinely does on its own without asking anyone's permission or explaining itself. Far from being responsible for damaging the earth's climate, civilization might not be able to forestall any of these terrible changes once the earth has decided to make them. Were the earth determined to freeze Canada again, for example, it's difficult to imagine doing anything except selling your real estate in Canada. If the earth decides to melt Greenland, it might be best to unload your property in Bangladesh.

Thus, the geologic record suggests that climate ought not to concern us too much when we're gazing into the energy future, not because it's unimportant, but because the energy crisis will be upon us before we succeed in changing the earth's heat balance in a major way. We know this because the changes to the climate caused by humans thus far are small compared to those known to have occurred naturally in the ancient past.

The geologic record also accounts for why the issues of energy and climate often seem so otherworldly. It's because the sixth mass extinction and the end of fossil fuels are events of geologic time. Like the coming of nuclear weapons or the industrial revolution, their approach is an imminent collision between the world of the imagination and the world of the here-and-now in which they will merge briefly and then separate again with altered definitions. Everything will come out fine, but it's scarcely surprising that conversations about both sometimes seem a bit psychotic.

CHAPTER 3

Jungle Law

W hen I was a kid, my father would always buy the cheapest gasoline. Not paying a penny more than necessary wasn't just a convenience for him. It was a matter of pride. He constantly kept his ear to the ground for deals and always knew who was getting in and out of the business and who was charging what at a given moment. If he caught wind of a slight price advantage somewhere across town, he would drive there immediately and fill up. If there were a gas war, he wouldn't just be happy but strangely cheery and hummy all the time, as though he'd gone to heaven but just hadn't yet picked up his harp.

My dad's obsession with cheap gas wasn't that unusual, of course. We learn this to some extent just by aging and observing our fellow citizens' buying habits. We learn it especially well, however, when we travel. A casual conversation about fuel headaches with a taxi driver in China, Europe, Latin America, or Africa, either directly or through an interpreter, always winds up being the same conversation. It's quite amazing. We discover that the person thinks exactly the way we do, right down to tiny details. Everybody wants the cheapest gas.

Once we understand how similar most other people in the world are to us, we begin having trouble with the concept of energy conservation for the good of all.[62] We want to be responsible, but our minds revolt and wickedly conjure up a guy in Bengal with a large family, a shiny new car, crushing education responsibilities for the kids, and ambitions to protect himself in old age by making more money. He pulls into his local gas station, starts pumping, and muses during the rather long wait, for the tank is capacious, how nice it is that people in Germany, France, and the United States are working so hard to use less gasoline, thus keeping prices low for him. He gets happier and happier as he pumps, and this is not just because of the lovely weather either, for it's hot and rainy. On his way to the kiosk to pay, he hums to himself in a contented sort of way, as though he were a carbon copy of my father, which of course he is. He then hops in the car and turns on the air conditioner full blast. Our minds then replicate this scene a billion times, in different languages and circumstances around the globe, each time playing out substantially the same way and for the same reasons. Then they ask for—and get—a small chuckle.

Energy's immense personal significance to all of us, revealed by the extremes to which we'll go to get deals, causes the law of the jungle, rather than the law of man, to regulate its production and use. Most of us don't think about this notorious fact of life very much because doing so isn't a good use of time. Energy is so cheap compared with other things we have to pay for—medical costs, education costs, mortgages, and so forth—that we're better off leaving gasoline injustice for someone else to worry about.[63] However, everybody understands that this cheapness is less a taming of the jungle than a fortuitous abundance of bananas. It's a consequence, in particular, of cutthroat competition for market share. When the competition goes awry, as it does from time to time, we find ourselves either paying what sellers demand for fuel or doing without. Someone else will buy if we don't. At such moments we are reminded of just how tough the energy business is and just how powerless governments can be when things get difficult.

The observation that energy production and use are fundamentally economic matters, not technological ones, isn't trivial, however, as hard-boiled Wall Street types are fond of claiming it is, because it powerfully affects what's likely to happen in the next century as fossil fuels run out. It implies, in particular, that a miraculous scientific discovery or invention is unlikely to have much influence on events. To do so, it would have to be both brilliant and economically on track, and that is an extremely tall order. Absent extraordinary and unprecedented intervention in the world's energy pricing machinery, it's difficult to imagine any future except marching in stages from the cheapest energy source, to the next cheapest, to the next cheapest after that, and so forth as resources deplete, presumably with considerable pricing pain at each transition. There isn't any way to stop this progression except to pay higher prices for gas, which none of us is willing to do. This is the environmentalist's nightmare scenario, of course, in that it consumes all the fossil fuels one by one and ends with coal, the dirtiest of them all. Such is life in the jungle.

The brutality of the energy business is so central to understanding the future and so difficult for most of us to grasp that it's worthwhile reviewing the history of a specific energy jungle incident in detail, just to get our facts of life straight. There's a long list of such incidents to choose from: the nineteenth-century consolidation of the oil business,[64] the roller-coaster fortunes of shale,[65] the intrigues of tar sands,[66] great coal strikes,[67] and Lady Thatcher's closing of the pits in Britain.[68] The optimal one, however, is arguably the California energy crisis of 2000–2001.[69] In addition to being contemporary and important, it's extremely well documented, thanks to press obsession with the story and some fortunate accidents that occurred as the crisis was unfolding. This enables us to piece together what happened on our own, independent of anyone else's analysis. It's also big enough to showcase the full spectrum of jungle behaviors, not just some of them. These include the drive of power generation and distribution industries to maximize profits, the drive of democratically elected governments to take those profits away and distribute them to the people, the drive of both buyers and sellers to manipulate government regulatory law through the political process to benefit themselves, and the struggle of everyone to take the moral high ground in public discourse and to define on their own terms what is right and good, even though what's actually at stake is money.

The crisis occurred in California, so naturally it had an air of zaniness that belied its seriousness. The 1996 Electric Utility Industry Restructuring Act, the immediate cause of the catastrophe, had been guided through the legislative process by state senator Stephen Peace, the famous creator, writer, and star of the cult film *Attack of the Killer Tomatoes*.[70] Soon after the law went into effect in 1998, signs began to appear that California was about to experience its own attack of killer tomatoes. The demand for electricity began to soar, as did the wholesale price of power. Shortages materialized. The state entered a period of rolling electricity blackouts that lasted a year.[71] One of the state's three major utilities tripled residential power rates. The other two utilities, constrained by law from raising rates, began borrowing billions to buy electricity on the spot market to cover customer demand at a loss. Both eventually went broke. One filed for Chapter 11 bankruptcy.[72] The other arranged a $3.3 billion bailout deal with the state. California governor Gray Davis declared a state of emergency and ordered the sale of electric power bonds and purchases of power for the grid by the state water commission, thereby committing $6 billion of state revenues to power supports. He justified these extreme actions on the grounds that he was defending California against the depredations of evil Texas energy companies.[73] He subsequently entered into $42 billion of long-term power delivery contracts with several energy companies (some of them from Texas), which were later criticized as being economically disadvantageous to the state.[74] Residential power rates climbed 62 percent. Thanks in part to the latter, he became the first governor of California, and the second governor in the history of the United States, to be recalled from office by popular vote. The voters replaced him with Arnold Schwarzenegger, the famous bodybuilder and star of the *Terminator* film series.

To be fair, the underlying cause of this problem was a flawed theory emanating from Washington. California simply had the misfortune of being the first state to comply with U.S. Federal Energy Regulatory Commission Order No. 888, a radical electric power restructuring concept that, it turned out, had bugs.[75] The most significant of these bugs was a rule forbidding long-term contracts between electricity producers and distributors. States that restructured later learned from California's mistakes, weakened this rule, and spared themselves most of the headaches.[76]

The government's objective in discouraging long-term contracts, thus forcing the various parties to shop at the last minute, had actually been to lower retail electricity prices. This is exactly backward from the way shopping usually works and seems especially incomprehensible given what happened when the plan was executed. It was, however, perfectly logical at the time. The premise of Order 888 had been that long-term prices were too high because the local power utilities were vertically integrated monopolies. If you wanted to lower costs, especially for the big industrial users, you thus had to force the utilities to divest their power generation capability. They would then purchase power for their customers from one of several power providers, who, in turn, would compete with each other in a newly created wholesale market. But you needed to disallow long-term contracts between utilities and providers when you did this, because such contracts would effectively reinstate the vertical integration you were trying to disrupt. Thus, reversal of the no-contract rule in other states that deregulated was a major course correction, not a minor one.

We know a great deal more about the inner workings of the California energy crisis than we might have because of its linkage to a vastly more wonderful and stupendous energy jungle incident, the collapse of the Enron Corporation.[77] Enron was a $100 billion multinational energy company specializing in the very things that concerned California—electric power delivery, natural gas delivery, and the building and management of power plants.[78]

From prosaic beginnings in 1985 it had rocketed to Wall Street superstardom, eventually being ranked by *Forbes* as America's most innovative company six years in a row. Whether induced by events in California or the poor performance of Enron stock, a number of people began to question Enron's opaque accounting practices and the strange fact that no one could figure out how it made money.[79] There followed a series of events that led, one year after California declared its state of emergency, to what was then the biggest bankruptcy in U.S. history.

The company's fall was precipitous and spectacular. Enron's stock, which had peaked in August 2000 at $90 per share, plummeted by July 2001 to $50 per share. Evidently not everyone was impressed by the company's claims that its second-quarter profits had increased a breathtaking 40 percent over the previous year's ($404 million from $289 million), that its electricity sales in North America had doubled, and that its electricity sales in Europe had increased fivefold.[80] On August 15 Enron's chief executive officer unexpectedly announced his resignation after only six months on the job, citing personal reasons.[81] On October 16 Enron announced that it would take a $1 billion charge against earnings and declare a loss of $618 million in the third quarter.[82] On October 31 the Securities and Exchange Commission opened an investigation into Enron.[83] On December 2, Enron filed Chapter 11 bankruptcy.[84]

Subsequent public scrutiny of Enron's records in conjunction with its bankruptcy proceedings revealed that the company had, in fact, contributed materially to the California energy crisis.[85] California officials had suspected as much during the crisis but had insufficient evidence to do anything except issue loutish threats.[86] Many of Enron's actions had been business-as-usual politics and thus, though distasteful in many people's eyes, not expressly illegal. For example, Enron head Ken Lay had direct ties to the White House and used them to lobby for two specific actions good for his company but hurtful to California: continued strengthening of Order No. 888 and denial of California's request for temporary interstate price caps.[87] However, other actions had been criminal. In May 2002 documents came to light that revealed Enron energy traders to have actively manipulated prices behind the scenes to exacerbate shortages and create additional arbitrage opportunities for themselves.[88] The precise amount of damage they did to California remains controversial, but the settlement eventually negotiated was $1.53 billion.[89]

There remains today no completely satisfactory account of exactly how California's rules led to shortages, how much the shortages were, and where the extra money California paid for power during this period (estimates range as high as $42 billion) went. When the law went into effect in April

1998, the average wholesale price of electricity was about $30 per megawatt-hour (5 cents per kilowatt-hour).[90] It stayed there until the crisis began in June 2000, when it abruptly rose to $150. At the time most people attributed the rise to an unhappy coincidence of strong economic growth, inadequate investment in power plants, hot weather, and drought in the Pacific Northwest. However, the price remained high through the fall and peaked at $330 in December 2000, a time when the weather was obviously not hot. It then hovered around $200 through the spring of 2001 and then abruptly fell in June. By September 2001 it was back to $45 per megawatt-hour, where it restabilized. By then California had a surplus of energy and had to sell some outside the state at a loss.[91]

However, subsequent events revealed that at least some of the shortage had been man-made. As investigations into Enron's electricity trading abuses proceeded in 2002, the spotlight fell on one Timothy N. Belden, a young, brilliant, and high-ranking officer in Enron's Portland electricity trading office.[92] Mr. Belden was unusually well informed about the electric power grid's vulnerabilities through his previous academic research career at the Lawrence Berkeley Laboratory, and he used this knowledge to implement scheduling schemes with colorful names like "Death Star," "Fat Boy," "Get Shorty," and "Ricochet."[93] These generated both large profits for Enron and wholesale price increases in California. Mr. Belden and two of his subordinates subsequently admitted these activities and pleaded guilty to conspiracy to commit wire fraud.[94] More importantly, the high public profile of their cases spurred the Federal Energy Regulatory Commission to pursue an unusually bare-knuckled investigation of human factors in the crisis. The Commission eventually found that not only Enron, but also nearly all the players in the California energy market, including the utilities themselves, had engaged in complex buying and selling practices that destabilized and increased prices.[95] The picture that emerged in 2003 was less a clandestine plot of an evil Texas company than a barroom brawl out of a low-budget Western. The Commission, evidently not amused, negotiated $6.3 billion in "monetary settlements," which is a polite regulatory way of saying "fines."[96]

The saloon in which most this brawling took place was the California Power Exchange (PX), a special energy spot market created by the 1998 law—and subsequently dissolved in January 2001 as an emergency response to the crisis. It had been created with an arcane mix of free-market principles and regulation that was, as it later turned out, almost perfect for honing the skills of young arbitragers. The 1998 law required producers and distributors to submit bids to the PX one day ahead of the proposed delivery date. The PX was to then formalize the sale at a clearing price, bundle agreements into a balanced (equal load and supply) dispatch plan, and forward this plan to the California Independent Service Operator (ISO) for scrutiny. The ISO, the agency responsible for managing the actual power flow, would then look for overloads at grid bottlenecks. If the submitted plan scheduled more power to flow through one of these bottlenecks than it could handle, the ISO would either call for additional bids so as to reduce traffic over the bottleneck or simply order reduction and compensate the affected parties with a fee. The congestion adjustments would then be bid on the day-of market at the PX just before the power was actually dispatched. As a safety measure against abuse of the congestion relief process, the day-of market price was capped. But the ISO had emergency authority to obtain power from outside the state at the going rate, whatever that was, if the amount of power available fell below the state's needs.

The undesirable behavior that surfaced first at the PX was price fixing. In its final report on the crisis, the Commission accused nine major power generation companies of initiating the crisis through "economic withholding" and "noncompetitive bidding."[97] It stopped short of claiming that the companies had colluded. It found the companies' rebuttals unpersuasive and inadequate to explain the dramatic price rises that began in May 2000. As evidence of uncompetitiveness it noted (1) that the spot electricity price had decreased slightly over the summer even as drought and heat conditions were worsening and the cost of natural gas was doubling and (2) the participants had continued to bid vigorously at the price cap as it was lowered over the summer from $750 per megawatt-hour to $500 and then to $250. None of the respondents denied that they had bid their marginal generation capacity way above cost. They only argued that it was their right to do so and that the Commission had thus not demonstrated that their action constituted withholding. The Commission countered that it wasn't their right to do so because the companies had functionally consented to certain behavior constraints when they chose to bid in the newly created market. It then instructed the companies to show cause why their behavior between May and September 2000 did not constitute a violation of regulations and grounds for disgorgement of unjust profits. There followed a steady stream of orders to return money to California.

The undesirable behavior that surfaced after price fixing was lying to the ISO. After the ISO price cap was reduced to $250 per megawatt-hour, the big utilities began deliberately underbidding their loads. It was a perfectly rational business decision. They had not yet divested all their generating capacity, so the government had, as a stopgap measure, required them to sell all this capacity into the PX day-ahead market and then buy it back at market rates. But the utilities quickly discovered that the day-of price was often lower than the day-ahead price because the sellers would be approaching a deadline with surpluses they had to unload. Accordingly, they began squeezing the sellers by buying less and less power in the day-ahead market and more and more power in the day-of market, a practice that made them temporarily net energy sellers and burdened the day-of market with enormous volumes of last-minute trading that it wasn't designed to handle.

The combination of deliberate load misrepresentations and last-minute shopping then caused prices to skyrocket in times of large demand. On days that the amount of power for sale in the day-of market wasn't adequate to cover buyer demand, the ISO would declare an electricity emergency and allow power to flow in from outside the state to cover the deficit. But this action effectively removed the price cap, because interstate rates, as opposed to California rates, weren't regulated. Naturally the out-of-state providers who came to the rescue at the last minute charged a fortune, often raising their asking price way above the still-regulated retail electricity price. Doing so presented the utilities with the terrible choice of instituting blackouts or buying power from scalpers and reselling it at a loss.

The undesirable behavior that surfaced after lying to the ISO was megawatt laundering. In times of tight supply in the capped day-of market, a California producer would deliberately sell its power to an out-of-state buyer, thus making supplies even tighter. Then, when the inevitable energy emergency was declared, the out-of-state partner would sell California back its own power at exorbitant uncapped rates.[98] Interstate rate caps could have stopped megawatt laundering in its tracks, but the Commission delayed ordering such caps until June 2001, presumably because it didn't completely understand what was happening. It was national policy at the time to deregulate the electric power industry at the wholesale level, and the Commission was understandably hesitant to contravene this policy.[99] The imposition of caps ended the crisis.

The undesirable behavior that surfaced after megawatt laundering was withholding power. With price caps high, as they were at the beginning of the crisis, or evaded, as they were later, it became profitable to worsen shortages by bringing generators down for "maintenance" during times of peak load. The FERC suspected this practice of being widespread and conducted a correspondingly wide investigation, but it found sufficient evidence to levy fines only in one case.[100]

The undesirable behavior that surfaced after withholding power was scamming the congestion-relief fees. The basic idea here was that companies would schedule extra fictitious traffic over congested bottlenecks and then have the ISO pay them not to send the fictitious traffic. This required some sophistication, because the dispatch had to consist of matched load and generating capacity, both fake. The company also had to make sure that its own fake traffic, not someone else's real traffic, got canceled. In one of the schemes, the celebrated "Death Star," the latter was achieved by scheduling power transmission in the opposite direction from the main flow in a congested line. This transmission was then balanced with an equal and opposite one through a line outside the ISO's jurisdiction. The ISO, unaware that the dispatch was really a giant, meaningless loop, would then cancel it and pay the corresponding fee. Meanwhile no electricity was ever transmitted, and no congestion was ever actually relieved. The "wheel-out" scheme was similar, except that it involved scheduling transmission through a line known to be out of service. The "load shift" strategy required controlling both generating capacity and load on both sides of a bottleneck. The company would send in a dispatch plan with load overstated on one side and understated on the other in the same amount, thus maintaining required global balance but also creating extra fictitious transmission over the bottleneck. It would then bid to remove this transmission, receive a fee, and restore the dispatch to what it should have been in the first place.

Finally, the undesirable behavior that surfaced after scamming the congestion-relief fees was shorting. The ISO required all power delivery to have contractually attached, as a kind of insurance policy, contingency generation and transmission capacity that could be called into service in case of emergency. These so-called ancillary services, sold as a commodity on the spot market, were cheaper on the day-of market, presumably due to everyone's attempts to unload them before time ran out. A company could thus make money selling ancillary services that it did not actually have in the day-ahead market and then purchasing generation capacity to cover its position the next day. This practice required lying to the ISO, however, because the rules required sellers of ancillary services to identify specific generation capacity.

The ultimate irony of the California energy crisis was that deregulation raised retail power rates rather than lowering them. This was true not only in California, which accepted long-term increases in exchange for restabilizing its market in the summer of 2001, but also in the seventeen other states that deregulated afterward. Such an outcome was perhaps not surprising given that arguments for lowering costs for everyone through restructuring

were always somewhat vague when it came to whose incomes would correspondingly decline and why the people taking the hits wouldn't mind losing
the money. It was, however, a welcome sanity check. The state suffering the
most was Texas, which deregulated shortly after California and experienced
60 percent retail price increases between 2002 and 2006. Texas regulators
even levied a $210 million fine against the big state utility, TXU, for physical
withholding.[101] The steady parade of price increases exhausted enthusiasm
for restructuring in the United States and eventually halted the implementation of Order 888. One state, Virginia, even re-regulated.

A sobering lesson we learn from incidents of this nature is that saving the
earth by reducing carbon burning is, in fact, a very low priority for most
people. California's electricity presently comes almost entirely from the burning of natural gas in turbines. During the crisis, we heard lots of talk about
corporate greed, market principles, needs of the elderly, socialist inefficiencies, and so forth, but we heard essentially nothing about generator efficiency,
alternate fuels, or slowing global warming by using less electricity—as Cali-
fornians definitely did during the crisis. Those things were not relevant. We
can reasonably expect this to be the case in the future as world energy supplies tighten.

Thus, green energy technologies, which by definition are more expensive
than non-green technologies, are inherently problematic. We can't have energy coexisting at two different prices at any given point in time in the jungle,
because clever arbitragers will always exploit, and thus eliminate, any such
differences. We can imagine complex taxation arrangements in which industries pay low prices for fuel while ordinary people pay high ones, thus enabling one's country to go green while its export businesses remain globally
competitive, but such arrangements are politically unstable because they're
unfair. The California example shows that even mild cost imbalances can radically alter the course of elections—and political careers.

Unfortunately, every one of us is responsible for this unhappy state of affairs, myself included. Like my dad, I tend to seek the cheapest gas, although
not as obsessively, for he was quite over the top when it came to cheap gas. I
suffer extreme remorse when I do so and endeavor to ride my bicycle instead
of driving whenever I can. But when something important comes up, such
as schooling for the kids or getting to a job interview on time, I'll burn fuel
with the best of them and not waste even one second feeling guilty. Billions
of my brothers and sisters around the world will be doing the same. We might
from time to time reflect on the irony that the short-term benefit to us, the
right to buy fuel at the cheapest price, enforces the law of the jungle and

thereby fixes the long-term destiny of fossil fuels, but we won't do so for very long because it's not our problem.

Meanwhile, demand for energy will grow, notwithstanding everyone's wish that we could stop this trend, slow the rat race, and relieve environmental stress on the earth. The reason is that a country's energy consumption correlates powerfully with its gross domestic product.[102] The exact size of this correlation is sensitive for climate diplomacy reasons, and also because energy use and economic activity aren't measured consistently throughout the world. However, the trend is extremely clear: The ratio of gross domestic product in dollars to energy consumed in joules is about five times the cost of electricity in dollars per joule.[103] If you don't understand math or don't trust charts and figures, you can just visit different countries and see for yourself. It's perfectly obvious that rich ones use more fuel than poor ones do. We don't know whether they use fuel because they're rich, or they're rich because they use fuel, but this is in some sense immaterial. Everybody with a clear head in the developing world—which is a great many people—sees this effect and understands that increased energy use means improved life for themselves and their families.

CHAPTER 4

Carbon
Forever

Centuries from now, when people no longer use fossil fuel, carbon will still be with us. A lot of it will be in the air causing mischief as the greenhouse gas carbon dioxide, a legacy of civilization's past excesses—at least until slow geologic processes return it to the earth—but the remainder will be circulating around the way it has always done, moving from the air into the bodies of plants, and from there to animals that eat them, and from there back again into the air, in an endless cycle. People will still be using carbon for things other than food as well—cotton for clothes, wood for furniture, and paper for a variety of things. We can even imagine them building bonfires now and then for fun, perhaps with leftover wood bought from local farmers needing to replant orchards. After lighting the thing off they'd sit back, gaze up at the cold night sky, and marvel at how the fire's heat was somehow just right. The kids would be roasting marshmallows near the coals and discovering how these flame up and blacken if they miscalculate, even though the sticks on which they're impaled don't. Mosquitoes would be plentiful, and there would be ants. We call the earth the water planet because it looks so beautiful and blue from space, but we really ought to call it the carbon planet, because that better describes what goes on down here.

Carbon is so central to the biosphere, including the maintenance of our oxygen atmosphere, it's perhaps not so surprising that carbon-based fuels like firewood and gasoline are magically appropriate in scale and weight for human needs.[104] Nuclear fuel rods contain a million times more energy than wood, but they're so dangerous, to both health and civil order, that they're

31

essentially useless for everyday applications. If jet fuel contained ten times less energy, we couldn't power airplanes with it because the weight would load the plane down too much. Carbon-based fuels are extremely light, both because carbon atoms are light and because they exploit the chemical activity of oxygen in the air instead of attempting to be self-reliant. They're also easy to control by changing the amount of air getting to the flame. Further, carbon atoms disappear into the air after use as carbon dioxide, a nonpoisonous gas essential to life and, better yet, destined for eventual capture and recycling by plants.[105]

It is thus a reasonable bet that people will still be using carbon-based machines in some capacity long after fossil sources run out. Carbon is just too superior in portability, lightness, and safety to all other energy storage media, at least here on earth. Where the energy to make the necessary fuels will come from is an interesting question, but it's somewhat secondary because there are several possibilities, and they aren't mutually exclusive.

Predicting the technological future is a notoriously dangerous undertaking, of course, so it's essential when doing so to take the concept of betting literally. Betting means that we put our own money down on the table with the understanding that we'll quadruple the amount if certain things happen and lose it all if they don't. The money in question must be ours, not someone else's, for the process to work properly. If we're likely to personally suffer from betting unwisely and losing, we might reasonably be leery of betting on, say, antigravity technology or perpetual motion machines because we'd have good reason to believe that they are impossible. We couldn't prove they were impossible, of course, because we never can prove things to be impossible, but we'd strongly suspect it and think better of placing the bet. We might also be leery of betting against well-established physical laws such as conservation of energy, thermodynamics, and quantum mechanics, even by accident. We could certainly imagine a world in which those laws didn't hold, but it would be unwise.

Betting may seem like an unnatural way to approach such a serious matter, but it has many benefits, among them sweeping away false authority. This is both crucial to credibility and consistent with genuine science, which is not about promoting learned opinions but rather about doubting things and checking them. This goes double for learned opinions based on no facts. We find the latter with increasing frequency as science gets more sophisticated, unfortunately, which is probably why hard-boiled business types, who know better, tend to disbelieve sophisticated scientific opinion and write it off as a ploy to usurp their authority and take their money. Dealing with this problem is exasperating from the point of view of physics, a notoriously precise and uncompromising science, but nothing is gained by arguing. We must, therefore, talk about facts so long as we can, and beyond that just bet seriously and see who wins.

The physical law most central to future fuel options, or lack of them, is quantum mechanics. This one deceptively simple principle underlies all of chemistry and is responsible for such disparate things as the atomic nature of matter, the size and shape of atoms, the tendency of atoms to bind together into molecules, the length of the bonds thus formed, and the amount of energy released or absorbed when molecular bonds rearrange in chemical reactions.[106] Quantum calculations require considerable work to execute properly and are not for the faint-hearted, but they can, with adequate time and a little help from a computer, predict the bond lengths and cohesive energies of substances we've never seen before to high accuracy. There is no doubt that quantum mechanics is correct.

One of the more interesting consequences of quantum mechanics is that all atoms and chemical bonds are roughly the same size—about one hundred-millionth of a centimeter.[107] This is so even though atoms differ enormously from one another in mass and chemical personality. A carbon atom is 12 times heavier than a hydrogen atom, a uranium atom 238 times heavier. Carbon atoms like to connect themselves to four other atoms, whereas hydrogen atoms like to connect themselves only to one other. One carbon atom combines with four hydrogen atoms in a tetrahedron-shaped molecule called methane, the principal component of natural gas. Hydrogen atoms left to themselves combine in pairs into rod-shaped molecules, the principal component of hydrogen gas. Uranium atoms are chemical troublemakers that bond in many ways and like to agglomerate into crystals in which each atom has twelve neighbors.[108] Nonetheless, the size scale of these various atoms and their bonds is more or less the same.

The energy stored in chemical bonds is also roughly the same.[109] This is easier to understand than it might seem, for it just means that the amount of tugging one must do to pull a bond apart is roughly the same. The bonding of atoms is analogous to the attraction of rubber balls with magnets buried inside them. If we bring such balls together, their mutual tug will increase until their surfaces touch, whereupon the rubber will begin to compress, push back, and lessen the tug. The equilibrium separation occurs when the ball squashes enough to resist the magnets. We can't get the balls closer together than this without pushing, because the elastic repulsion overwhelms the attraction at small separations. Bond energies are always negative, meaning that atoms like to bond. We get energy out only if we increase the number of bonds or if we trade weak bonds for stronger ones. Most burning is in the latter category.

The springy repulsion of atoms at close separations makes liquid and solid fuels devilishly difficult to compress. This is obvious to anyone who has pumped gasoline, but the numbers are nonetheless educational.[110] If we compressed gasoline to, say, twice its normal density, we would have twice as many atoms in the tank and thus twice the chemical energy stored there. This would be advantageous, say, for increased driving range per tank. But we would also have energy stored in the compression itself, and this is not a mere detail. The springiness in question, which is set by quantum mechanics, is so stiff that this extra energy would amount to one-third of the combustive energy stored in the gas before we compressed it.[111] Thus our tank would hold two and a half times the energy it did before, not two. However, the pressure required to achieve this compression would be about one million atmospheres, the same as that created by high explosives when they detonate.[112] Obviously, no metallic tank could contain such pressure, and even if it could, we wouldn't want such a tank because an accidental structural failure would amount to setting off two hundred sticks of dynamite in the car.[113]

Thus, how much energy a fuel can pack away in a small space, something central to its function, is fixed by quantum mechanics, not serendipity. It isn't an accident that carbon-based fuels carry the amount of energy they do. It's more or less the maximum energy density we can ever get with any substance without resorting to millions of atmospheres of pressure. Not only is it impossible to increase gasoline's energy content using processing tricks, it's also impossible to do it by substituting other atoms for carbon in the fuel. At best we'll get exactly the same energy content, and at worst we'll get degraded content and, as an added benefit, a poisonous combustion product that the environment can't neutralize easily.[114]

This packing issue is also why we can't eliminate the carbon. Conventional gasoline and diesel fuel contain hydrogen, the other substance that burns in air and leaves behind an environmentally friendly combustion product: ordinary water. We might find ourselves contemplating the use of pure hydrogen as a fuel instead of gasoline, thus reducing air pollution. But we would soon discover that hydrogen is a damn nuisance. It's an explosive gas under earthly conditions, it doesn't store easily, and it's expensive.[115] Indeed, at the moment the only cost-effective way to make hydrogen is by removing the carbon from carbon-based fuel, thereby making less energy, not more.[116] These problems come about because hydrogen atoms can only bond to each other in pairs, so they're fundamental. In other respects, though, hydrogen is nearly as good a fuel as gasoline. A given tank filled with highly compressed hydrogen gas contains about one-tenth the energy it would contain if filled with gasoline and about one-third that filled with cryogenic liquid hydrogen.[117] However, both storage methods are highly dangerous, and, more to the point, inferior to the conventional chemistry of hydrocarbon fuels. The carbon in gasoline, in effect, remediates hydrogen's design flaws by compressing it to liquid densities using quantum mechanics.[118]

Another thing that quantum mechanics regulates, the tendency of chemical bonds to pop apart when pulled, similarly limits a material's ability to sustain tension, or negative pressure, and thus limits the energy we can store in high-pressure air tanks, flywheels, and springs.[119] The maximum energy per unit mass we can store by any of these techniques is roughly the apparatus material's strength-to-weight ratio, a number several hundred times smaller than the energy density of gasoline.[120] This disparity exasperates many engineers, who believe they could make materials with high enough yield strengths to challenge gasoline if only they could eliminate the material's atomic imperfections.[121] Their proof of principle is the diamond anvil cell, a tiny high-pressure tank made of defect-free single-crystal diamonds that routinely exhibits yield strengths two hundred times greater than those of steel.[122] But even if we could control a material's imperfections (a tall order when we're making things with heat and chemical reactions), it might not matter in the end because tanks or flywheels made of such materials would be vastly more dangerous than tanks and flywheels that exist now, which are already exceedingly dangerous.[123]

Quantum mechanics is also responsible for the strange coincidences we find when we compare one energy storage method with another. For example, suppose we cool a conventional gas, such as air, to its liquefaction point and put the liquid in an open tank, as we might with gasoline. We could then extract energy by warming the liquid back up to ambient temperature and running a steam engine with the high-pressure gas that boils off.[124] The energy would actually be coming from the environment's heat, but this is an unimportant detail. However, no matter which working fluid we consider, air or something else, the energy stored per unit mass always comes out suspiciously close to the strength-to-weight ratio of optimal tank materials.[125] This makes the energy storage capacity of cryogenic liquids functionally the same as that of compressed air, regardless of which liquid we choose, even though the two things appear unrelated at first. We might hope to overcome this problem with a gas that liquefies at room temperature but relatively low pressure, which would allow us to store it at high density in tanks that don't weigh very much. Alas, gases, such as carbon dioxide, that liquefy at low pressures also deliver correspondingly little energy when vaporized, so the energy stored per unit mass comes out again to be the same.[126]

The most curious of these quantum coincidences is the similarity of compressed air energy densities to those of batteries.[127] It isn't enormously surprising that batteries should contain roughly as much energy as other storage media, the energy source of a battery being the same as that of a gasoline engine or a campfire—the rearrangement of chemical bonds. The forces drawing atoms together to form bonds are electric in nature, and batteries cleverly exploit this fact to produce electricity directly instead of producing heat first. This saves us the bother of using engines and generators to power flashlights. But the surprising thing is that batteries have exactly the same weight problem that air tanks do. In theory they should be better storage media than gasoline, not worse, because they're optimally efficient and generate no waste heat. However, they lose this efficiency advantage through the extra weight they must carry in order to function. Conventional batteries don't use oxygen from the air but instead carry all their reactants on board. These reactants tend to be much heavier than carbon, hydrogen, and oxygen for performance reasons (the one exception being lithium, which is one of the reasons battery people love lithium).[128] Batteries also carry their own waste products because the latter can't be exhausted as vapors. But the truly important thing is that batteries contain nonactive materials that control the chemical reactions, channel the energy produced by them into electricity, and conduct the electricity out to the user. No matter how hard we try to reduce this extra weight, it always manages to add up to one hundred times the equivalent amount of gasoline, making batteries no better as energy storage media than compressed air.[129]

It seems counterintuitive that something as organic as natural gas combustion could be electric in nature, but it's quite true. We can, with appropriate (and expensive) materials, construct fuel cells that convert methane and atmospheric oxygen directly into electricity without generating heat.[130] It's easier to make this conversion if we use hydrogen as fuel instead of methane, but hydrogen isn't essential.[131] In fact, the fuel cells of greatest interest to researchers at the moment burn methanol, a fuel that is liquid at room temperature and thus easy to store in tanks.[132] Replacing batteries with such fuel cells would have large and important lifetime and weight reduction benefits in consumer electronics and military robotics. For similar reasons, there is great interest in air batteries, which are electric storage devices halfway between fuel cells and batteries that discharge oxygen into the air and then take it up again as they work, similar to the way plants do.[133]

Nonetheless, no matter how sophisticated fuel cells and air batteries get, they'll never pose a serious challenge to gasoline. In fact, the argument goes the wrong way. If fuel cells advanced sufficiently to burn hydrocarbons, such as methane, at low cost, they would, if anything, increase gasoline's lead over its competitors by removing its noise, pollution, and waste heat disadvantages while leaving intact all of its advantages. But they'd never do the reverse. A more likely scenario, however, is that fuel cells will remain too expensive to have any effect one way or the other. The problem is that they all use platinum, a precious metal more expensive than gold. The need for this scarce material is not a mere coincidence but actually a consequence of quantum mechanics, for the special chemical inertness that makes platinum essential for fuel cells also makes it rare in the earth's crust and correspondingly difficult to find. Obviously, platinum's scarcity would become more serious as fuel cell use grew. Thus, with hefty government subsidies we might imagine having a few locomotives, buses, and forklifts powered by fuel cells, but without subsidies we probably won't have even these things, because cost-conscious buyers (which is to say, most buyers) will choose conventional engines, which are simple, reliable, and cheap.[134]

The great quality and effectiveness of conventional internal combustion engines powerfully reinforces carbon's hegemony. We don't normally think much about our car's motor, and rightly we shouldn't. It functions so well that fretting about its inner workings is a foolish waste of time. But it's actually a remarkable piece of machinery that takes up little space, delivers enormous power for its weight, and costs very little to make.[135] It's admittedly a big hunk of metal, but this is deceptive because metal is cheap and because the engine is delivering five hundred times the power for its weight that a horse does— or a human being.[136] If we're willing to pay a little more money, we can get a gas turbine, which delivers five *thousand* times the power for its weight that a horse does.[137] The cost and weight advantages of turbines are crucial to commercial aviation, which simply could not exist if the only available power sources were batteries or compressed air tanks. They're also important for the electric power industry, which uses gas turbine plants almost exclusively to cover its load peaks.[138] But one doesn't have to be an expert on jets and electric power plants to see the benefits of internal combustion engines. Everyday experience with cars tells us that they're marvelous things, and that their low procurement and maintenance costs are central to the economic calculus of energy.

Thus, when the oil begins to run out a few decades from now, the world will almost certainly respond by replacing conventional petroleum distillates with synthetic fuels that act like them and are used in exactly the same ways. Batteries, hydrogen fuel cells, and other such things will not be relevant. Part of the reason is that hydrocarbon fuels are optimal, in a physical sense, and so could be displaced only by inferior products that customers wouldn't like. But an equally important reason is that retooling the entire world economy would cost more than retooling the world's oil refineries. The latter cost is known with some precision because the requisite technologies already exist.[139] It's roughly the cost of rebuilding the world's refinery fleet from scratch. Synthetic fuel plants are not widely deployed at the moment because conventional petroleum distillates have a slight price advantage, but that advantage will disappear when petroleum begins to run out.

Another reason to anticipate that synthetic fuel will prevail is that the retooling probably won't actually cost the industry anything. Past experience tells us that such costs easily get passed on to customers—which is to say, get repudiated through inflation because that's what passing energy costs on to customers means in practice. The trick behind this amazing inflationary vanishing act is that fuel costs tend to ripple through the economy and slowly manifest themselves in the prices of everything else.[140] This is why gasoline always seems to have the same effect on our lives no matter how much its price goes up.[141] It's presumably also why the fraction of the U.S. gross domestic product dedicated to energy hovers around 10 percent, and why energy consumption is more or less the same multiple of gross domestic product around the world.[142]

The development of a synthetic fuel industry is blocked at the moment by green politics, not merely by price competition from petroleum. The grounds for opposing it are that synfuel manufacture is a huge carbon dioxide emitter. The central problem is synfuel's hydrogen component, which must be obtained either by reforming of natural gas or by brute-force breakup of water. The latter requires enormous amounts of energy.[143] This energy doesn't have to come from burning coal, but if it does, as is normally the case for cost reasons, then the manufacture and use of synthetic fuel winds up putting twice as much carbon into the air as conventional petroleum distillates would have—by some accounts six times more. The carbon dioxide problem doesn't occur if the hydrogen comes from natural gas, but then one is functionally in the business of turning natural gas into gasoline and diesel fuel, not replacing natural gas and petroleum with something else. The synthesis process itself is so cost-effective that converting natural gas to these other fuels makes economic sense right now as a means of exploiting the market price difference between them. (Gasoline and diesel fuel are also easier than natural gas to handle and transport in ships.) Accordingly, there is much industry activity at the moment in gas-to-liquids fuel synthesis, as the natural gas version of synfuel is called.[144] Manufacture of synthetic fuel from coal, by contrast, is done commercially only by South Africa's Sasol, the plant in question being a legacy of apartheid.[145] Environmentalists routinely vilify this plant as one of the world's worst carbon hot spots and point to it as the reason we don't want a synthetic fuel industry, even in countries that have lots of coal and growing petroleum import burdens.[146]

Nonetheless, the first source of the carbon in the synthetic fuels that eventually replace oil and natural gas is likely to be coal. This is a horrific prediction from the environmental perspective, for it would commit the world to putting every last atom of burnable carbon presently in the ground into the air—and at accelerated rates. It's conceivable that people of that time will be sufficiently convinced that the globe is warming and sufficiently alarmed over the consequences that they will eschew coal and get their carbon from somewhere else. However, the most reasonable guess is that they won't, and that they will rationalize coal burning the way we do today, even if the world is slightly warmer, and continue the practice out of expediency. Coal is currently the cheapest source of carbon in the world by a large margin, and its price advantage over its competitors will only increase as oil and natural gas deplete.[147]

The coal will last only a few decades, however, and as it exhausts, it will have to be replaced with something else. By far the easiest solution to this problem, and thus the one people are likely to adopt, is to substitute biomass—agricultural products of various kinds—for coal as feedstock in the synthetic fuel plants already built.[148] It's actually possible to do this today, since the precise nature of the feedstock is immaterial to fuel synthesis, but we don't because coal is so cheap. Of course, the substitution would stress the world's agricultural system terribly and amount to going backward historically. (Deforestation was one of the major factors leading from wood burning to coal burning in sixteenth-century Europe.[149]) But by this time the original grounds for green opposition to synthetic fuels would have evaporated, for any carbon released into the air during manufacture or use of the fuels would have been extracted from the air in the first place by the plants and so would be greenhouse-neutral. Thus, in a wonderful ironic twist, the final outcome of the move to synthetic fuel is likely to be the renewable and green carbon budget that most environmental groups long for today, even though they commonly oppose subsidies for synfuels on the grounds that synfuels aren't green.

The central issue at the time of this substitution won't be energy procurement but carbon procurement. Carbon and energy are effectively synonymous at present, but this is only because coal and natural gas are so cheap that we can profitably burn them to produce electricity. That, in turn, makes the cost of electricity higher than the cost of these fuels. But once the world's carbon is all in the air, this situation will reverse. Retrieving the carbon will require large amounts of energy—indeed more than the carbon yielded when it burned—and this will make carbon-based fuels more expensive than energy generally. The initial price disparity will be set by the biomass itself, which yields more energy when simply burned to make electricity than it does when converted to gasoline and diesel fuel.[150] That will cause the price of electricity to fall below the price of these fuels. Whether it drops further depends on the production costs of electricity from alternative sources, such as solar cell farms, geothermal mines, or nuclear reactors. If these costs are high, then the alternative sources will give way to cheap electricity made from biomass, and the price disparity will remain fixed. But if, as is more likely, these costs are low, then biomass electricity will give way instead to the alternatives, and the price disparity will grow. The result will be a two-tier energy industry in which biomass supplies carbon-based fuel while alternative sources supply electricity at prices much lower than those of the fuels.

Conventional agriculture (including farming of the oceans) will remain the key source of carbon for these fuels no matter how cheap electricity gets and how clever high technology becomes.[151] One can imagine scientists of the future finding a way to extract the necessary carbon out of the air artificially at low cost, thus making agriculture irrelevant to the world's carbon supply, but one should snap out of this daydream as soon as possible. Green plants have been extracting atmospheric carbon (and hydrogen) for a living for three billion years, and they simply aren't going to be out-engineered by a bunch of scientists. Moreover, one of the themes we see repeating again and again in the energy business is that deferring large capital investments as long as possible or, better yet, avoiding them altogether and shifting the responsibility and risk onto someone else, is highly desirable. This calculus strongly favors farming over factories. Agriculture isn't tidy, and it involves lots of hard work, but it requires relatively little upfront capital and has the grower, rather than the processor, doing the borrowing. Agriculture even avoids facility costs completely, in a way, because the plants build their own facilities using marvelous methods that people would love to emulate but can't. Thus, very basic economic considerations suggest that the fuel business will involve agriculture no matter what technical innovations come out of the laboratory.

Where all this energy-related agriculture is going to take place is an interesting question. We already know that even minimal diversion of good land and water resources to biofuel crops raises food prices, creating severe political problems as it does so.[152] Diversion of sufficient resources to supplant the entire present-day geological carbon budget would have a devastating effect on the world's food supply. It's conceivable that a new economic equilibrium will develop in which most good land is reserved for food while the rest is allocated for industrial carbon—which is then made into carbon-based fuels that fetch sky-high prices because they are dear. A more likely possibility is that a new carbon supply industry based on saltwater agriculture will develop.[153] There has been much speculation about how and where such an agriculture might be built, for it's an obvious long-term solution to the food competition problem. Lots of plants grow well in saltwater, perhaps even the bulk of the plant matter on earth. Open oceans or shoreline estuaries might be suitable locations for great farms, although one has to be careful there about damage to fisheries. Experience with irrigated crops in arid climates tells us that deserts would be great for growing such crops, for they have high temperatures (which plants love), lots of sunshine, and few clouds. At a time when people are talking openly about powering all of Europe with solar energy from the Sahara, it might be profitable to also think about irrigating part of the Sahara with saltwater to grow algae there. We then repeat this calculation for other parts of the world with deficiencies of freshwater, such as Patagonia, Western Australia, and Northern Mexico, and imagine the day that they become vibrant and prosperous places because of the carbon agriculture taking place there. However, the world will have to wait for the coal to run out to learn how this drama ends, for the agricultural product in question is functionally a proxy for coal, and the laws of economics will prevent any such agriculture from developing so long as coal is plentiful and cheap.

Thus, the end of fossil fuel burning that occurs two centuries or so from now will be a carbon revolution, not an energy revolution. People will still use the same kinds of transport and heating technologies that they do today, but the mix will be different on account of changes in pricing. An important new industry, the production of biomass feedstocks for fuel production, will have come into existence and will be competing economically for land and water with the food industry. Everybody will be paying through the nose for gas. The present carbon dioxide buildup in the air will have ceased, as will further degradation of the environment from the greenhouse effect. But, although the economy will be founded firmly on green plants, it will probably not be green. The economic forces encouraging the use of nonbiological materials that don't degrade will still be present, and there will still be population pressure and its corresponding problems of pollution, deforestation, and declining biodiversity. Unlike the carbon problem, these things won't solve themselves.

CHAPTER 5

Pipes of Power

While the world's carbon problem is busy solving itself, the world's electricity industry will be coming of age. Electricity today is sophisticated and central to everything that happens in a modern economy, but it is nonetheless still dependent on fossil fuels for its existence on account of being made from them, rather like a child who is old enough to leave home but won't. But all that will change when fossil fuels begin to exhaust. Electricity will continue to be sought after as a consumer product through the transition, and it will continue to be a superb medium for transmitting nonfossil energy of various kinds in a universal form to its buyers. Thus, at some point in the future, the price per unit of electric energy—more precisely, the price per unit of nonfossil energy delivered electrically—will cross below the price per unit of fossil energy. When that happens it will be like a pond turning over in the fall. Things that used to be expensive will now be cheap and vice versa, and all those who use significant amounts of energy will be scrambling to rethink their business models. Electricity will then acquire a life of its own and slowly supplant its parent as the dominant energy industry.

It's difficult to overstate how marvelous electricity is. You flip a switch and a huge room lights up as though several large bonfires had winked into existence. The lights elsewhere in your neighborhood dim together imperceptibly as you do this, then return to their original state as, somewhere far away, the governors of enormous machines you've never seen, and probably never will, sense the extra load and pour on extra steam or fuel to supply the needed power. Your refrigerator hums quietly in the kitchen as a small motor, in a reversal of what's happening at the generating plant, consumes electricity to turn an axle connected to coolant pumps. It stops humming occasionally as a little brain inside, also powered by electricity, senses the temperature and turns the motor off, and then on again, so that the food inside remains cool enough not to spoil but warm enough not to freeze. Another, somewhat larger, brain in the next room is projecting dancing images onto a screen faster than your mind can assimilate them. Meanwhile downtown, industrial-strength motors are raising drawbridges, cooling office buildings, and lifting elevators half a kilometer into the sky—a lucky thing, considering how useless buildings this tall would be if we had to climb their stairs. Huge pumps are transporting rivers of fresh drinking water into the city and rivers of sewage out. And as night falls, the city transforms into something unimaginable without electricity: a vibrant commercial world of motion and neon illuminated with the light of five billion candles.[154]

Despite its remarkable versatility, electricity is not fundamentally different from any other kind of energy. Electric wires are simply a transmission system, like the one in a car that goes from the engine to the wheels. More precisely, they're like the hydraulic lines of a backhoe or skip loader. In such machines, compressed oil, driven by pumps, surges through pipes to the control levers, and from there to the cylinders that move the great steel arms and buckets at your command. In the electric grid, a compressed electron fluid, driven by generators, surges through wires to switches, and from there to the motors that move drawbridges and elevators at your command. We have volts instead of hydraulic pressure and amps instead of oil flow, but otherwise it's just the same.

The great miracle that makes electric power transmission possible is not the phenomenon of electricity itself, which is rather commonplace, but the special ability of metals to confine electricity to certain paths and lead it to specific destinations. The degree of this ability is breathtaking. A piece of aluminum, for example, is one billion trillion times better at conducting electricity than is a piece of polyethylene the same size.[155] This remarkable property is quantum mechanical and characteristic of all metals. It's the same thing that makes them shiny, ductile, and strong, properties desirable for jewelry and absolutely indispensable for building bridges, skyscrapers, and engines. It's part of their birthright as metals, as it were. Metals are rare in nature, at least in their native state, because all of them are vulnerable to corrosion (rusting). Oxygen in the air loves to complete chemical bonds, and it attacks metals and leaves behind a crumbly, useless mineral that has none of the metal's special properties. Thus, although electricity itself is not man-made, the metals required to make it useful definitely are.

The conductive property of metals enables wires made from them to function like pipes. When we apply a pressure difference between one end of a garden hose and another—for example, by opening the valve at the house end—water already in the hose speeds up and eventually flows out the open end at a steady rate that depends on the pressure difference and any constrictions. These constrictions include the hose itself, which obviously transmits more water if it's short and wide than if it's long and thin. Metallic wires behave just like this. They carry a steady electric current that depends on the voltage difference between the two ends and any constrictions. The latter include the size and length of the wire itself, which transmits more electricity if it's short and wide than if it's long and thin. The conduit must of course have a boundary through which fluid can't pass. In the case of the hose, it's the plastic from which the walls are made. In the case of the wire, it's either plain air or a coating of plastic, resin, or ceramic.

What flows through this pipe when it transmits power is a swarm of electrons, the ghostly subatomic particles present in all matter and responsible for strength, density, and chemical bonding—and, of course, lightning and midwinter doorknob shocks. In insulators, which is to say, in most materials, the electrons are tightly attached to their atoms and can't move. But in metals they're only loosely attached, and they slip quantum mechanically from one atom to the next, thereby becoming mobile and fluid.

However, electric current is much more than the wanderings of a few unhappy particles led astray by quantum mechanics. Electrons, unlike atoms, are clothed in light. More precisely, electrons have surrounding them a kind of halo so intimately connected with radio, microwaves, light, and so forth that it's functionally indistinguishable from them. This halo is visible as the familiar repulsion of a pair of party balloons that have been rubbed on one's head. The connection with light is easy to see. We just imagine one of the balloons to be whisked away suddenly. The other balloon cannot know that its brother is gone right away, because news of the whisking cannot travel faster than the speed of light. Accordingly, the news must arrive one light delay-time later at the earliest. It does so in the form of a radio pulse or, if the whisking is sufficiently violent, a visible light pulse.

The electromagnetic field, as the electron's halo is called, is the true agency of electric power transmission. Were only one or two electrons involved, this would not be the case, for the fields would then simply be telegraphing to distant observers what the electrons were doing at any given moment. But the number of electrons in a wire is stupendously large—about twenty billion trillion per cubic centimeter—and this causes the overlapping halos to take on a life of their own and dictate what the electrons do, rather than the other way around. Under the right circumstances they can even jump off the line and propagate away as pure radio, leaving the electrons behind. As a consequence, it's perfectly correct to think of electric power transmission as a radio wave, traveling at the speed of light, channeled to propagate along a certain path by wires but otherwise unaffected by them. That's why it doesn't matter what metal the transmission line is made of.

An important consequence of this intimate relationship between electricity and light is that a long transmission line's operating voltage fixes its power-carrying capacity. For the highest voltage presently used in North America, this capacity is about 2 billion watts.[156] This contrasts sharply with transmission over very short distances, such as in a house's wiring, where the power delivered is smaller and is limited only by things such as load and wire thickness. But in the long-distance case, the electric and magnetic fields always have a certain special relationship with each other by virtue of being a trapped radio wave, so the voltage determines everything.[157] In turn, the ability of insulators to confine the electric pressure in the wires limits the operating voltage. Because the pressure in question applies only to the electrons, a high-voltage wire is never in danger of exploding into shrapnel as a high-pressure tank might be. It is, however, in danger of exploding out blasts of electrons. The buzzing and crackling we hear when we pass under high tension lines is due to corona discharge, a benign precursor to full-blown lightning. Somehow we sense that mighty forces are at work in a wire when we hear this sound, and we're right. The electron pressure in question, over twenty billion atmospheres, is deliberately engineered to be near catastrophic flashover—within reasonable safety margins, of course.[158]

Because electricity pipelines have limited capacity, we can profitably compare them with oil and gas pipelines. The amount of power transmitted by one of the three wires of a big line is roughly the power consumed by a jumbo jet at full throttle on its takeoff roll.[159] This fact inadvertently answers a question that has perplexed all air travelers since the dawn of aviation, which is how such a massive machine could possibly fly. The answer is that it requires the power of a small city. The fuel consumption equivalent to this power is about thirty times the flow rate of gasoline into a car's gas tank when it's being filled up.[160] Transporting it would thus require a hose about six times thicker than a filling station pump hose. The power delivered by the entire transmission line is equivalent to three times this flow, or ninety times the pumping rate of gasoline into a car. A typical natural gas pipeline, which has a diameter of about one meter, delivers about ten times this power.[161] A major oil pipeline, which also has a diameter of about one meter, delivers about one hundred times this power.[162] Thus, although the transmission line is a pipe, it is not an extraordinarily large one.

We can also profitably compare the flow of energy in a large transmission line to the flow of energy in a large river. More precisely, we can compare it with the power of a large waterfall, either a natural one or an artificial one made with a dam, because it makes no sense just to take energy out of a flat river, such as the tidal Amazon. The river would stop flowing if we did, causing the water to pile up and preventing us from extracting any more energy. We find that a large transmission line carries slightly more power than Victoria or Iguazu Falls, half the power of Niagara Falls (before they were tapped to make electricity), one-third the power of Guaira Falls (before they were drowned by the Itaipu Dam), and one-thirtieth the power of the Congo River's Livingstone Falls, the largest falls on earth.[163] It also carries three times the power of Hoover Dam, slightly less than the power of the Aswan High Dam, half the power of the Grand Coulee Dam, and one-tenth the power of the Three Gorges Dam, presently the largest dam on earth.[164] Thus, although a big transmission line carries the power of a very large river, it doesn't carry the power of the largest ones. But no river carries more energy than a big oil pipeline.

We can also profitably convert the flow of energy in a large transmission line into flow of money. This is perhaps the most useful conversion of all, for the seductive beauty of great rivers tends to fool us into overestimating the value of the energy they supply. At the retail price of electricity appropriate to a developed country, the flow through a big transmission line amounts to about $40 per second.[165] This is approximately five times the amount a large traffic bridge takes every second in tolls,[166] or the amount three hundred McDonald's restaurants take in every second from their hamburger and milkshake sales.[167] It is also the interest on ten billion dollars.[168] The latter is obviously a serious constraint on what we can and cannot do with transmission lines. There is, for example, no technical reason why overland transmission lines have to be conspicuous and ugly. We could, in principle, just bury them. However, if we did, we'd never recoup the cost of the burial, plus maintenance, because the market value of the energy we'd be delivering is just too low. Thus we leave the lines open.

Open transmission lines generate measurable effects on the ground that are notorious, both as a public health concern and as something that probably shouldn't be a public health concern.[169] Electricity pipelines become literally pipes if we surround them with metal tubes. We do this, for example, with undersea power cables, where leakage of the electric and magnetic fields into the saltwater would result in serious losses (and also great unhappiness for fish).[170] Sheathing a transmission line in this way has no effect on its power-carrying capacity except in lowering the voltage it can sustain, which comes mostly from the wires being closer together. The electric and magnetic fields are equally trapped by the open lines of overland tower transmission, but they decrease rapidly with distance from the line rather than stopping abruptly at a boundary.[171] The residual fields on the ground under the tower, the ones that affect people, are not extraordinarily large. The maximum electric field strength is the same as that delivered to our head by a party balloon after we've rubbed it and stuck it to the ceiling above us.[172] The maximum magnetic field strength is one-fifth that generated by the earth, the familiar orienter of compass needles.

There is no fundamental technical limit to the distance over which transmission lines can pipe power. At the moment lines are rarely longer than one thousand kilometers, but this is mainly due to cost constraints. One thousand kilometers of overland transmission line is about what we can build with a billion dollars.[173] A line longer than that would lose money, because operating costs plus service on the debt would begin exceeding the market value of the power it's delivering. But if money were no object, we could easily stretch power lines across continents or, for that matter, between them under the seas. We'd have to use special cables for the latter, but except for that, there's no difference from piping power across land. Indeed, undersea power transmission over distances of less than five hundred kilometers is now routine, particularly in Europe, where short saltwater barriers are especially plentiful, but also in Japan and Australia. Money is always an object, unfortunately, which is why longer-distance projects—such as the link between Iceland and the United Kingdom, which is perfectly feasible technically—have so far remained dreams.[174] A similar fate may be in store for newer long-distance power transmission schemes such as the energy superhighway system and the European offshore supergrid, although time will tell.[175]

A big transmission line also doesn't leak significantly. Were its resistive losses large, they would prevent us from transmitting power over very long distances the same way that holes in a garden hose prevent us from watering the far reaches of our lawns. But a meter of transmission wire generates only about as much heat as your body generates sitting in a chair.[176] It's warm to the touch for this reason. However, this power pales in comparison to what's coursing down the line. The total power that a thousand-kilometer line loses to heating is about one-tenth of the amount it carries, or four dollars per second. We could easily lower this loss by making the wires fatter, but we don't normally do that because the interest burden on the extra aluminum we'd have to purchase would exceed the value of the power saved.

The marvelous capabilities of electric transmission lines are, unfortunately, very good at hiding where the power comes from. For most of us, the energy comes from the transmission line, and that's that. Worrying about how it gets there is like worrying about how food gets into supermarkets. But how it gets there is central to properly assessing the energy future, because it is energy supply, not energy transport, that will change fundamentally in the next two centuries. At the moment, about 70 percent of the world's electricity comes from burning coal and natural gas.[177] We don't see the generation plants, but they're there, way on the other end of the transmission line, pumping more carbon dioxide into the air than we would if we just burned the fuel locally, thanks to inefficiencies of generation and transmission. This fraction is likely to increase over the coming decades, not decrease, as renewable energy advocates wish it would, because demand will be growing, particularly in developing countries, where people will be understandably unenthusiastic about paying extra for electricity made from the wind, sun, and tides. Thus, these pipelines of power are, as a practical matter, facilitators of fossil fuel consumption, not substitutes for it.[178]

Fossil fuel is especially central to present-day electric power because it readily meets marginal demand. When we flip on a light switch, we want the light to come on right now, not several hours from now when the power company finds it convenient. This instant response is so important to us that we will switch power companies, or make political unhappiness for the company we have, if it's not provided. If we're an industrial user, we might even get fed up and relocate our company to a place that doesn't have this problem. As far as we're concerned, brownouts and blackouts are their fault, not ours. But we can have power on demand only if the power company has spare generating capacity that it can turn on whenever we require. Wind and solar plants don't have this property because they generate power only when the wind blows and the sun shines. Murphy's Law guarantees that this will never be at a time convenient for us. Hydroelectric facilities can—and do—increase the water flow to their turbines in times of increased demand, but unfortunately this doesn't help much because only a small fraction of the world's generating capacity is hydroelectric. What's left are coal and natural gas plants, which can be turned up or down easily by burning more or less fuel. Electric power providers like to say that they manufacture electricity when it is needed, but it might be more accurate to say that they use carbon-based fuels to store the energy until it is needed, just as cars and trucks do, and for the same reasons.

The demand problem is considerably more serious than a few light switches in a house, however. The peak demand on a typical workday, which occurs in late afternoon, is 50 percent bigger than the minimum demand just before daybreak.[179] For a large city such as New York or Chicago, this difference amounts to two billion watts, or the entire capacity of a large transmission line. For the geographical region in which the city resides, a more appropriate unit for assessing this surge effect, the number is more like twenty billion watts. The peak is also higher on especially hot days, when everyone wants their air conditioners on. There are also large seasonal variations. Thus, the important demand variations aren't individual caprices but enormous ebbs and flows akin to tides. Accommodating peak demands while minimizing the cost of idled generating capacity during slack time is extremely difficult and, arguably, a power company's core business problem.

Demand variations wouldn't be such a problem if electric energy could be banked, but unfortunately it can't, at least in its native form. Electricity's light-speed propagation also makes it ethereal. The flow of energy through a big transmission line may power a city, but the amount of energy contained in the line at any one time, even when the line is very long, is less than that in a small wine glass full of gasoline.[180] If the generators at one end of the line were to shut off suddenly, this energy would propagate all the way to the other end of the line and drain out in less time than it takes a nerve impulse to propagate from your fingertips to your brain.[181]

Also, both electricity-banking and transmission-line capacity are circumscribed by quantum mechanics. It imposes electric and magnetic strength limitations on materials analogous to their physical strength limitations. For example, if we subject a good insulator, such as glass or plastic, to an electric field even one-thousandth of an atom's internal electric field, the material will discharge in a mighty burst of internal lightning and blow itself up.[182] This weakness, which is analogous to the rupturing of a tank if we pressurize it too much, limits the energy density in and around any electric storage apparatus to about one-millionth that of gasoline.[183] The situation with magnetic storage is similar, except that we need a perfect electrical conductor to make it work rather than a perfect electrical insulator. This is surprisingly easy to achieve, for we merely need to cool a conventional metal down to ultralow temperatures.[184] But a sufficiently large magnetic field destroys this perfect conductivity, and this then limits the magnetic storage density we can achieve. It is considerably greater than the electric density but still only one-hundredth that of gasoline.[185]

The largest amount of magnetic storage in the world at present is at the Large Hadron Collider, a huge particle accelerator that resides in a tunnel twenty-seven kilometers long under the city of Geneva.[186] It stores the energy equivalent of a fully loaded jumbo jet flying at cruising speed.[187] Unfortunately, this is also the amount of energy a big transmission line delivers in nine seconds, the retail value of which is $356.[188] The accelerator cost $9 billion to build.

Our limited ability to store electric energy in its native form is probably just as well, for trapped electric and magnetic fields sufficiently large to be useful would also be extraordinarily dangerous. Not only would they have the energy density of gasoline—which is to say, dynamite—they would explode at the speed of light if the apparatus confining them failed. The amount of energy we'd have to store to accommodate the daily demand variations of a large city is also staggeringly large, about one-half of a small atomic bomb.[189] Moreover, the danger would be synchronized with the sun, so it wouldn't occur in just one big city but rather in every big city in all the world every day, a total of two hundred atomic bombs accumulated and released again every twenty-four hours.[190]

We can bank electric energy by converting it to some other form of energy and storing that, but the corresponding technologies are all bedeviled by issues of safety, environmental impact, efficiency, and cost. None is completely adequate. Forms that will work include mechanical energy (flywheels), stress energy (compressed air), thermal energy (hot salt), chemical energy (batteries), and gravitational potential energy (water pumped uphill).

The best of these energy storage technologies at the moment, and the only one currently deployed on a large scale, is water pumping. It seems astonishing that such a simple thing as water could solve a great problem like energy storage, but it's quite true. Before gas turbine plants became cheap and efficient, hydropumping was the method of choice for smoothing out peaks and troughs in daily demand. There are presently twenty billion watts of hydropumping capacity in the United States, enough to handle about a fifth of its daily surge, and slightly more in Europe.[191] Hydropumping facilities buy electricity off the grid at night when the price is low, pump water uphill with it, wait until the next day when the price is high, and then release the water back down through their penstocks to generate electricity for sale. One loses about a fifth of the energy when one does this, but the price difference is so great that one still makes money. However, a hydropumping facility has capital costs about twice those of a gas turbine delivering the same peaking power, and they're additionally burdened with large land acquisition and legal costs, not to mention the headache of dealing with government. As a result, they're economically problematic so long as natural gas is cheap.[192]

Thus it's the economics of storage as much as the economics of transmission or production that locks the electricity industry into fossil-fuel burning. This problem is classically illustrated by the case of Japan, a country lacking domestic carbon fuel resources of any kind and that would love, for balance-of-trade reasons, to convert its electric grid entirely to nuclear power. However, it can't at the moment because nuclear plants can't economically meet the surge demand. Japan thus covers its base load with nuclear power and accommodates the surge with peaking plants fueled with imported natural gas.

Unfortunately, the basic physical and economic constraints on the electricity industry are so fundamental that they're unlikely to be mitigated by any technological advance that occurs in the next two centuries as fossil fuel resources deplete. Electric transmission lines are marvelous vehicles for transporting energy, but they'll never be anything more. They'll never get much bigger capacities than they have today, and they'll never be able to store appreciable amounts of energy in their wires. There will always need to be a generator at the other end of the line that manufactures electricity instantly from some other source, and the core issue of electric power supply will always be what that source is. There might conceivably be cosmetic improvement, such as laws requiring ugly transmission lines to be underground, and there might conceivably be more lines, perhaps even under the oceans, but none of these things will affect the economics of electricity production and use at all. People will still demand that electric power serve them, rather than the other way around, and that will still obligate the ultimate energy source, whatever it is, to have enormous reserves on hand that it can bring to bear the instant they're required, just as fossil fuel plants do today.

Accordingly, we can predict with some confidence how exhaustion of fossil fuels will affect the electricity industry. Carbon-based fuels will still be available, but they'll be extremely expensive and thus no longer cost-competitive with other sources, notably nuclear energy. Nuclear energy won't necessarily be what replaces fossil-fuel burning, but it will establish a price ceiling below which any other energy source will have to fall to survive. The terrible danger of nuclear waste won't have gone away, but it's difficult to imagine anyone voting to save the earth from nuclear waste when doing so would greatly raise their electricity bills. In addition, surge capacity will be supplied with storage, as opposed to carbon-fuel burning in peaking plants, simply because it will be cheaper to satisfy demand that way than either burning carbon or adding extra plant capacity. Hydropump won't necessarily be what supplies this storage, but it will likewise provide a price ceiling below which any future storage technology will have to fall to survive. Thus the industry's final pricing structure, regardless of how much trauma it suffers while the coal is exhausting, is constrained to be quite reasonable by technical elements that exist today.

The creation of great storage capacity, in turn, regardless of its original purpose, will fundamentally alter the economics of energy from the sun and wind. Any storage system that can bank enough energy from nuclear reactors to meet the world's surge demand can also bank enough energy from solar and wind farms to even out their generation caprices. Renewable energy will thus become more competitive than it is today, regardless of whether nuclear energy arrives on the scene first. But whether nuclear energy or its renewable competitors will win out in the end is difficult to predict, for at the moment government intervention so muddles the true costs of both that we can't tell exactly what they are.[193] It will probably be a fight to the death, though, for there can be only one cheapest energy source.

Although hydropumping may not be the storage technology that prevails, it will set a very high performance benchmark that any competitor will have to beat. The small-scale contenders that exist today, such as batteries and compressed air that we pump into underground cavities, just won't scale up to the sizes we need to handle the surge of the world.[194] Hydropumping operates at about 80 percent efficiency, which no engine presently in existence can do, including the best steam turbines.[195] It's capable of reversibly absorbing as much energy as we could imagine throwing at it, much more than the world's entire nuclear weapons arsenal at its Cold War peak.[196] It can deliver energy back on demand at huge flow rates, and at a moment's notice. Its major mechanical part, the water being pumped, gets renewed and replaced every cycle, just the way air in automobile cylinders does, and so it never wears out. The turbine blades and shaft are simple, robust, and inexpensive to maintain over their forty-year lifespan.[197] The technology has an excellent safety record. It's also within our means—water being the most plentiful substance on the planet and free. All we need is a dam and a few electric turbines, and we're in business. If a newly hatched high technology achieved even one of these wonderful things, it would be a headline-making event, so high tech's achieving all of them simultaneously a second time, and to a degree that beats hydropumping in the marketplace, will be difficult.

The amount of hydropump storage required to make renewable energy a reality is large but not absurdly so. For example, when Lake Mead is full, its upper meter of water stores forty times the daily surge requirement of New York City.[198] That's also half the surge requirement of the United States and one-eighth the requirement of the world. Thus what's required to handle the world's daily surge is about eight Lake Meads. If we're willing to deplete the whole reservoir, not just the upper meter, we can store the entire world's electricity supply for a month in about sixty Lake Meads.[199] The total area of these lakes would amount to about 2 percent of Mexico or 0.5 percent of Canada or the Sahara.[200]

However, it's best not to get too carried away thinking about this mighty fleet of reservoirs or the glorious future ahead for electricity generally. None of this is likely to happen before carbon becomes depleted. It's just too expensive compared to burning ultracheap natural gas in turbines.[201]

CHAPTER 6

Inspiring Mammoths

N uclear energy is always a challenge to talk about because it's otherworldly. Most people know what coal is, how coal burning works, what kind of stuff goes up the smokestack when we burn coal, and so forth, so talking about coal burning is easy, albeit a bit exasperating to the listener, who usually has better things to do than discuss coal. Fundamentally it's fire, and everybody knows what fire is. But nuclear energy isn't an everyday thing, even though conceptually it's fire too, and it's immensely powerful and dangerous, so conversations about it tend to become metaphorical, which is never good in a technical matter. Nobody talks any more about nuclear-powered vacuum cleaners that never need plugging in or nuclear-powered cars that don't need gas, although they once did, nor do they talk about what kind of fallout shelter to dig and what kind of provisions to stock there.[202] Everyone understands that those things are misinformed nuttiness of the past. They do, however, talk about ridding the world of nuclear energy because of its transcendent dangers.[203] Nuclear energy is one of those things we're for or against the way we're for or against political parties.[204] Nuclear energy proponents are probably in the minority in the world at the moment, especially given that one of the biggest nuclear accidents in history, Fukushima Daiichi, has just taken place. There has been some talk recently (less so after Fukushima) of a nuclear renaissance,[205] and China is even now madly building nuclear power plants.[206] But if you live in the United States and want to impress your neighbors with how eccentric you are, just get a regular subscription to *Nuclear Plant Journal* or *Neutron News* and make sure it sticks out of the mailbox.

Unfortunately, thinking seriously about the future requires us to be very objective about this sensitive subject, for nuclear energy is going to be a major factor in the events that play out a few generations hence, when fossil fuel finally exhausts, no matter what. People might get electric power from nuclear reactors or they might get it from cheaper sources, but they'll never eschew cheap nuclear energy just because doing so is good. Cheap energy will still be a matter of great importance to them, and they'll put their own needs first, just as we do today. If necessary, they'll just invent reasons why those moralistic people of the past were misguided and why using nuclear energy was the right thing to do all along. The world that's coming will not be nuclear-free, although exactly how much nuclear energy it will have is unclear and dependent on events to come.

Nuclear reactions present two serious problems that set them apart from all other energy sources: The waste products they leave behind remain dangerously radioactive for very long times (one thousand years to three hundred thousand years, depending on your danger tolerance), and the explosions they facilitate are a million times more powerful than those of dynamite.[207] Both require extraordinary measures to mitigate, and both are intractable. No country has managed to commit itself to any long-term burial site for its nuclear waste,[208] even though everyone has understood the need for such burial for over fifty years.[209] No country has credibly quantified how good its security is at preventing determined people from diverting nuclear material from peaceful power programs to make weapons.[210] No country treats nuclear energy in a nonmilitary way. All countries censor news about nuclear energy—although some are better at it than others.[211] The underlying problems are so great that this situation is probably permanent.

The immediate health threats nuclear energy poses are scary but surmountable. The main poisonous effect is not from chemistry, as would be the case with cyanide or arsenic, but from subatomic particles emitted from the material at great speed. These are the notorious alpha, beta, and gamma rays of radioactivity.[212] Both they and the occasional neutron that comes out can penetrate through walls to get us, even though we can't see their source. When they do, it can be ghastly, for they disrupt a body's machinery in ways that can, in large doses, kill a person in two weeks or, in smaller ones, do it over the course of a few years through cancer.[213] They can also damage one's germ cells, thereby affecting one's progeny.[214] Fortunately, they don't penetrate very far. The technology for protecting oneself from them is no more complicated than burying the radioactive material underground or enclosing it in thick coats of concrete. The emitted radiation does, in fact, infect the enclosing substance and make it radioactive too, but this effect is confined to the immediate vicinity of the source and doesn't propagate. It's not the flu. Each successive infection uses up an enormous amount of energy, and there just isn't enough to keep the chain going. Moreover, these particular materials drink up nuclear energy like sponges and give back none, so they're terrific at shielding us from radiation.

More problematic is the span of geologic time that the waste requires to lose its radioactivity and become benign. The worst danger decays away in a few hundred years—a reasonable amount of time for human institutions to keep the waste sequestered. But no one can guarantee that something will remain buried in the same place for thirty thousand years. The proposition is absurd. Thirty thousand years ago, Canada and northern Europe were covered with ice. This was long before recorded history or, for that matter, before writing had been invented. Between then and now, enough rain fell on the world to fill up the oceans ten times. We'd be expecting a thousand future generations to mend cracks, clean up spills, prevent contact of the waste with ground water, and so forth, when we can't even get our kids to clean up their rooms.

The terrorism threat, both from nuclear explosions and from radioactive waste hijacked for use as dirty bombs, may also be surmountable, although no one really knows. It's ghoulish to talk matter-of-factly about this danger, but we must when we're being serious about nuclear energy, because it isn't zero, and it increases as we build more and better reactors.[215] The danger in question is very serious and probably amounts to the loss of a major city now and then, perhaps one in which one's own descendants are living. Moreover, we know from accidents such as those at Three Mile Island, Chernobyl, and Fukushima that nuclear trouble is big trouble, even if there's no fireball.[216]

Nonetheless, we can mentally count the number of major nuclear mishaps that would have to occur in the next two hundred years to dissuade people in desperate need of power from turning to nuclear energy when the coal runs out. A Chernobyl-sized event every fifty years would probably not suffice. A Hiroshima-sized event might not either.

Nuclear fuel is fossil fuel, and it's limited just the way coal, oil, and natural gas are. It's much more ancient, of course, for it antecedes the earth and is probably the handiwork of violent stellar explosions that occurred in the early universe.[217] Regardless of where it comes from, however, its supply is finite and will exhaust over time just as supplies of carbon-based fuels will. With current uranium consumption practices, the exhaustion will happen soon— between one hundred and two hundred years from now.[218] The total amount of easily fissionable energy in the ground is about one-third the amount of coal energy in the ground.[219] Thus, present-day nuclear power technology cannot substitute for coal burning in the long term, even in principle.

However, by the time the coal runs out the world will probably have extended the useful life of the nuclear fuel resources in the ground to about twenty thousand years by embracing breeding practices.[220] This phenomenal time increase, which is a bit hard to believe at first, is scientifically sound and backed up by experience with nuclear reactors, but it's very tough to achieve, for both technical and political reasons. The production-scale fast breeder reactors built in the United States, Russia, France, and Japan have been notorious for their huge cost overruns and mechanical problems.[221] They've also been lightning rods for antinuclear wrath and sabotage.[222] However, the most serious problem with breeding is that it would, if adopted widely, enormously ease the making of nuclear weapons—for everybody.[223] As a consequence, all governments have dragged their feet at developing full-blown breeding technology, and many have mothballed existing programs.[224] But the risk calculus will be very different when the world's coal is nearing its end. No government will oppose breeding if the alternative is no electricity.

The devil's bargain we make with efficient use of uranium comes from chemistry, oddly enough. Like most substances, uranium is a mixture of isotopes—atoms with the same chemical properties but slightly different masses. The energy content of these isotopes is the same, but it's only easy to extract energy from the rare one, U-235. To make nuclear weapons from uranium, we must therefore separate out this rare isotope. That is extremely difficult to do, for no conventional chemical process will work. We must use sophisticated technologies such as differential diffusion through porous media, centrifugation, or laser separation. Building and maintaining machines capable of doing this is a major undertaking and thus normally a great barrier to weapons construction. But all reactors, breeders or not, slowly convert the plentiful isotope U-238 to the transuranic element plutonium by a process called neutron capture. Plutonium, like U-235, has energy that's easy to extract for making electricity or weapons. Indeed, about one-third of the energy from a conventional light-water reactor comes from plutonium created by neutron capture and then burned on the spot.[225] But plutonium is chemically different from uranium and thus easy to extract and concentrate using ordinary wet chemistry. Thus any used reactor fuel is a potential weapons threat by virtue of the unburned plutonium it contains. Breeding maximizes this threat because it produces as much plutonium as theoretically possible—indeed, more than the U-235 (or plutonium) it burns—so the amount of useful fuel actually goes up. Thus, after we run these reactors a long time, the world becomes awash with highly dangerous plutonium. But the very effect that endangers us also keeps the lights on for thousands of years because it turns the otherwise useless U-238 into valuable fuel.

Efficient breeding is difficult to achieve technically because it requires running the reactor hot. The likelihood that a neutron will make plutonium increases with its speed, so the trick in breeding is simply to prevent neutrons freshly created from a nuclear burning event from slowing down too soon. However, conventional light-water reactors slow down neutrons quickly on purpose. That's why they contain water. All isotopes that burn in reactors have something called a fission resonance that makes it more likely to absorb a neutron and release energy if the neutron is very slow. Thus we can (and do) build an extra safety margin into our reactors by making sure that they just barely sustain a chain reaction if the neutrons are slow. Then, if the reactor accidentally loses its water coolant, as happened at both Three Mile Island and Fukushima, the neutrons will no longer be slowed down, and the chain reaction will stop. But efficient use of uranium requires us to abandon this time-tested safety principle, for designs complying with it can never achieve a breeding ratio greater than one. Therefore we must abandon water as the coolant. Most modern breeders use liquid sodium, a substance that conducts heat well and doesn't slow down neutrons, but unfortunately it explodes spontaneously into flame if it comes in contact with air. Most of the technical problems with breeders, at least the reported ones, have been sodium leaks.

The adoption of breeding actually has the potential to make nuclear fuel last considerably longer than twenty thousand years. A lot of people have noticed that the amount of uranium in seawater is only three million times less than the amount of salt.[226] This may seem like a minuscule concentration, but it's rather large from a chemistry perspective and something we can feasibly extract with ion-exchange techniques.[227] In fact, during the Cold War several governments built classified pilot projects to investigate how the costs of such extraction would scale with size. The cost of uranium obtained this way is difficult to judge accurately because of government involvement, but it appears to be about ten times the current world price of mined uranium. Thus, it's definitely crazy to extract uranium from the oceans for reactor fuel right now, but it might not be so crazy when conventional supplies begin to run low, particularly because breeding would make uranium one hundred times more valuable than it is right now. Thus, events unfolding when the coal runs out have the potential to turn the economics of seawater extraction on its head.[228] If uranium extraction from the sea were to became profitable, it would stay profitable until supplies were drawn down substantially, because the chemistry involved is elementary and not strongly dependent on concentration. That means most of the uranium in the sea would be available for harvest. That's a stupendous amount of energy, enough to supply all the power presently used by civilization for five hundred thousand years.[229] Moreover, the time in question is so long (the rains would fill up the oceans one hundred times) that geologic processes such as erosion and diffusion out of ocean sediments might very well replenish the uranium as it's extracted and prevent supplies in the sea from being drawn down at all. Were that to happen, the uranium could last a billion years, longer than life has existed on earth.

Fusion, the power source of the sun and a key player in the largest nuclear weapons, will very likely have been tamed for peaceful purposes by the time the coal runs out. This is a fairly safe prediction, notwithstanding the many times in the past fifty years promises of fusion just around the corner have turned out to be false. There are many indications at the moment that a key fusion event—laser ignition of a capsule the size of a coarse sand grain—actually is imminent, but old hands are reserving judgment until the ignition actually occurs. There will, of course, be several decades to deal with unexpected glitches. Like uranium isotopes, the necessary hydrogen isotopes are plentiful in the sea and could supply the world with energy for a billion years.[230]

Fusion is vastly more difficult than fission to harness for peaceful purposes because fusion is aptly analogous to fire.[231] Fusion works by building nuclei up from isotopes of hydrogen, just the way a chemical explosion builds up carbon dioxide from carbon and oxygen. Thus, what controls the rate of fusion isn't the supply of free neutrons, but rather pressure and temperature, especially the latter. Once a fusion reaction ignites, it tends to run away to completion and thus to cause an explosion. Indeed, one serious approach to controlling fusion at the moment is to limit the amount of fuel and just contain the explosion we get.[232] The other approach is to prevent explosion by holding the temperature rock steady at just the right value, the way the sun does.[233] The engineering challenges are formidable in either case because the temperatures and pressures required are otherworldly. In a conventional fission reactor, by contrast, the temperature and pressure are those of an ordinary steam engine, so they're quite manageable.

The central difficulty of inertial confinement, as the first approach to fusion is called, is making a small amount of fuel burn. Small samples are much harder to light off than big ones because their reaction products tend to escape without hitting anything and, therefore, without depositing enough of their energy to sustain the fire. To explode a capsule, we must thus first compress it to roughly a million times its normal density so that its reaction products are sure to hit something on their way out. After that, we must heat it to temperatures comparable to those in the sun's interior. Chemical high explosives don't supply nearly enough pressure to achieve either of these things, so we use laser light—a great deal of it. The latest big fusion laser occupies a warehouse the size of three football fields and delivers the energy of about a stick of dynamite to the capsule surface in a billionth of a second.[234] The time scale of the flash is so short that the power delivered briefly exceeds the power generated by all the world.[235] When and if the capsule ignites, it will explode with the energy of one hundred sticks of dynamite and make the room dangerously radioactive for a week.[236]

The central difficulty of magnetic confinement, the second approach to fusion, is holding the atoms in place. The sun doesn't have any such difficulty because it's a big gravitating body. If the fusion reactions in the sun's core begin to get out of hand, the extra heat they generate puffs the sun up a little, lowers the temperature in the interior through gravity effects, and slows the reaction rate back down. But the earth doesn't have enough gravity to perform this trick (which is why it isn't a sun), and a human-sized fusion reactor has even less gravity, so it can't copy the sun either. Instead we must emulate the sun's gravity with magnetic bottles.[237] Unfortunately, these bottles have led to one maddening engineering frustration after another, for confining a hot plasma with magnetism, it turns out, is like squeezing a bag full of cats.[238] Magnetic confinement is thus behind inertial confinement in the race to achieve useful energy technology. However, as Yogi Berra observed, it ain't over 'til it's over.

However, even if fusion is tamed, it is unlikely to rid the world of either fission or the dangerous nuclear waste that fission generates. The reason is simply that we can get about ten times more energy if we use the fast neutron the fusion reaction throws off to initiate fast-neutron fission and breed fissionable fuel.[239] Fusion enthusiasts love to talk about extracting nuclear energy without making any dirty fission products as a side effect, but they're just kidding themselves. No government or commercial enterprise would throw away 90 percent of its energy to save the world from nuclear waste. Moreover, it's easy to make this fission enhancement happen. We just surround the fusion chamber with a thick blanket of U-238 or thorium. The neutrons that escape the chamber (nearly all the neutrons) then plow into

the blanket and initiate nuclear reactions there. The great speed of these neutrons enables them to fission the blanket atoms with fairly high probability,
which they couldn't if they were slow. That already gives us a significant energy enhancement, for a fission reaction yields about twelve times the energy
a fusion reaction does. But the amplification doesn't stop there, for the blanket fission then produces between two and three daughter neutrons, which
then join the neutrons knocked out of blanket atoms with no accompanying
fission, plus the fusion neutrons that just slowed down without doing anything, in a soup. This soup then gets absorbed by the blanket atoms and turns
them into fuel suitable for use in conventional fission reactors. Thus, for all
practical purposes, fusion is not a clean alternative to conventional nuclear
energy at all but an alternative breeding technology, one that will compete
with breeder reactors for market share. It's difficult to predict at the moment
which will be cheaper, but obviously it can be only one.

Nuclear energy is not only likely to be cheap when the coal runs out, but
also dirt cheap. The reason is that low electricity prices are one of the few
tools we have for minimizing nuclear weapons dangers once we've resolved
to use fission for generating electricity rather than ban it outright.[240] The
problem is not so futuristic, actually, for it comes up today in the world's increasingly dysfunctional conversation about reprocessing fuel rods.[241] Cheap
electricity discourages weapons manufacture chiefly by removing the profit
from constructing a vertically integrated nuclear fuel industry. It's theoretically possible for any country to develop a nuclear fuel from scratch, thus ac-
quiring the ability to make nuclear weapons from scratch. Doing so is
extremely expensive, however, and it also isn't a very good use of money, in
that it induces the enemy, whomever that is, to build his own nuclear capability, which then sits on the shelf, as does yours, because neither of you wants
nuclear war. But if one floods the world with cheap fuel rods, thus enabling
anyone to make as much nuclear electricity as they like for almost nothing,
it becomes a money-loser to make fuel from scratch, just as it's a money-
loser today to make one's own gasoline from scratch. Cutthroat pricing would
also consolidate the industry, just as it consolidated the petroleum industry
long ago, thus making it easier to regulate. The chief regulatory tool would
then be something called return-to-sender, the requirement that rods be returned to the manufacturer after their power is exhausted. Anyone who failed
to account for their rods or returned them with a little plutonium missing
would then be denied further access to cheap rods. The other tool would be
spiking the rods with non-fissile isotopes, so that reprocessed fuel rods
couldn't be used for making weapons without expensive isotope separation

investments.[242] But neither tool could work even in principle unless there were a large entry barrier to fissile fuel manufacturing, which in turn would require the cost of fuel rods, and thus the price of electricity generated from them, to be extremely low.

Consequently, the end of coal is likely to unleash a chain of events that permanently locks electricity prices at very low levels, in the process making market penetration by fully green competitors, such as sun and wind, exceedingly difficult. The situation is likely to be similar to today's except with the price ceiling maintained by nuclear fuel instead of coal and natural gas. This so-called plutonium economy is a nightmare scenario for many people and a specific target of much contemporary antinuclear political activity.[243] However, unless the world rids itself of nuclear technology altogether, which is highly unlikely, nuclear power will remain in the background, disavowed by elected governments but nonetheless standing by, ready to expand into the economic vacuum left as coal and oil retreat. Thus, the plutonium economy may be inevitable.

Nuclear waste disposal will still be an enormous problem two centuries from now, but people will likely have accepted by then that nuclear power is necessary for living and that we must responsibly take care of the waste it generates, just as we take care of sewage.[244] Ignoring the problem in either case creates a potentially deadly health hazard, but dealing with it firmly isn't difficult and results in everyone being safe.

Although the adoption of breeding won't eliminate the nuclear waste problem, it will lessen it considerably. The bulk of a spent fuel rod, which is unused U-238, won't be waste any more but rather valuable feedstock and thus something people want to chemically isolate and recycle. Also, the transuranic elements other than plutonium, the chief source of radioactivity on scales greater than a thousand years, all fission in the presence of fast neutrons and thus could be mitigated and possibly burned up altogether by recycling.[245] Whether people would actually do this, however, or build fast reactors optimized for burning up actinides, depends on cost and how much long-term radioactivity they could tolerate in their waste, both of which are difficult to predict. But the remaining 0.5 percent of the original fuel rod that consisted of fission products would still be red hot and still require sequestration for a thousand years.

The amount of fission-product waste that a plutonium economy generates is shockingly small. A thousand-year accumulation of the stuff would occupy a cube about one football field on each side.[246] If people decided to remove and burn all their transuranics, the amount of waste in the ground would climb slowly to this value over a thousand years and then stabilize there, as people withdrew cooled-off material and substituted fresh discharges from their reactors. What would happen to the waste after its thousand-year slumber depends on the industrial value of the various substances it contained, but the easiest thing would be to just dissolve it in acid, take the solution out to sea somewhere, and spread it about, as you might spread ashes.[247] The dilution would reduce the intensity of any remaining radiation to less than that occurring naturally in seawater, both from dissolved salts and from the sky, and it would be a threat no more.

Because pure fission-product waste has relatively little military significance, disposing of it amounts to getting rid of garbage that nobody wants in their backyard. It will therefore probably get shipped across international borders to desolate places where few people live, the economy is terrible, and locals will welcome the opportunity to get paid for guarding the vaults. The Canadian high arctic comes to mind (although usually to the minds of Canadians), as does the Antarctic peninsula.[248] But the Russians probably have the best imaginable place—the truly awful region near Magadan on the Sea of Okhotsk, where the worst Stalinist labor camps used to be.[249] The waste could happily rest in peace there for a thousand years in deep mines, inspiring the mammoths. We can even imagine businesses springing up on the Road of Bones with fanciful names like "Dima's Hot Spot," "The Rare Earth," and "Actinide's." The alternative would be to deal with waste the way people deal with garbage now—by pretending it isn't a problem and then quietly paying private companies to bury it somewhere, no questions asked. That would, of course, pose a serious threat to underground water supplies, just as landfills do today, only worse because of the time scales involved. The Russians would probably be involved in that solution also, through their mafia.[250]

Desolate places might also wind up being where people locate their great breeding facilities. In principle these facilities could be sited anywhere, but in practice it might be more cost-effective to put them in places that antinuclear activists (or terrorist guerrillas) couldn't reach easily. Fast reactors will always be more dangerous than light-water reactors, so they'll always be unwelcome in most communities. But there would be no domestic security need to locate breeders nearby, or even in your own country, for that matter. They could very easily operate in some godforsaken place, cranking out fuel rods and shipping the copious amounts of electricity they produced as a by-product out on transmission lines to users far away. There are even some proposals to locate them under the oceans, although that's probably going too far.

But regardless of how people deal with this problem, or how things play out in the next two hundred years, such exploitation of nuclear energy as occurs won't threaten the long-term health of the earth. Nuclear phenomena often seem strange, but they're actually no more unnatural than blue sky or birds. Radioactivity has pervaded the environment since the birth of the world, and living things have learned how to heal the wounds it inflicts, so long as they're not excessive, just the way they heal bruises, cuts, and broken bones. Nuclear energy is admittedly terrifyingly powerful, but then so are coal, oil, the Livingstone Falls, and every other energy source capable of sustaining civilization. We just have to deal with them.

CHAPTER 7
Calling All Cows

I t had to happen eventually. Even the most desolate desert has an oasis, the most exhausted mine a fresh seam of gold. The gloom of global warming guilt had to break momentarily and let a ray of hope shine through. When it did, armies of frustrated journalists rose to the occasion:[251] *Pig Poo to Help Solve Energy Crisis! Manure Power Goes Live in Texas! It's Clean, It's Green, It's Cow Manure Power! Cow Dung Lights Up L'Oreal Factory! Dairy Cow Manure to Power Children's Train Ride! Tons of Turkey Dung to Help Fuel Power Plant in Minnesota! Dutch Harvest Chicken Poo to Power Ninety Thousand Homes! Plenty O'Poultry Power! Put a Chicken in Your Tank!*

Saving the world with manure is such a humor gold mine that it's delightful to think about day or night, but it's also serious, in that microbe-generated fuel is likely to become important two hundred years from now when the coal runs out. People probably won't be driving to work using gas made from cow pies because there will never be enough cow pies, but they are likely to be driving on gas with a cow pie component. At the very least, sky-high prices for carbon-based fuels will induce farmers to digest their manure and market the gas it generates rather than simply letting the manure decompose. The world may or may not go nuclear, and it may or may not have sound government, but it will always have plenty of poo.

There's a lot of energy in manure. This isn't surprising, for there's a lot of energy in the forage, silage, and grain the animals eat before they poop. If collected, dried in the sun, and burned, the world's agricultural manure would supply about one-third as much energy as all the coal presently consumed.[252]

If anaerobically digested first, it would render up about one-third of this energy as methane, enough to supplant one-fifth of the world's present natural gas consumption.[253] What digestion left behind, about half the original energy content, would normally become fertilizer, but it could equally well be dried and burned to produce, on average, a trillion watts, or about half the present electric power consumption of the world.[254] Although manure power is funny, it definitely ain't hay.

The immensity of poop's energy content is more than merely impressive, however, for it also tells us that agriculture has the capacity to supply the carbon part of the world's energy budget once the coal, oil, and natural gas run out. Manure itself would provide only 12 percent, even if we could collect it all (which we can't), but manure is just a waste product sloughed off by an industry dedicated to food production. We could easily raise the fraction if we made energy production our top priority. For example, we could double it just by removing farm animals from the picture and burning the crops directly.[255] We wouldn't want to do that, of course, because we need the food. The price structure of present-day animal methane, the stuff generating all the hilarious headlines, also tells us that agriculturally produced fuels will be reasonably priced. They'll be substantially more expensive than present-day fuels, but probably not ten times more expensive.

However, there's no getting around the fact that agricultural fuel production competes with food production for resources. This occurs to some extent even when we burn manure, because doing so reduces the amount of nitrogen-rich fertilizer available for spreading under food crops. But the earth can only render up so much energy from the good land it has, and the amount of this energy we would have to divert to power civilization is extremely large. For example, suppose we dried the entire U.S. corn crop, including stems and leaves, and burned it to make electricity instead of feeding it to livestock. This is an admittedly brutish way of getting the energy out, but it provides the cleanest accounting because all other extraction methods discard some of the energy in processing. This desperately reckless act, which would zero out the economy of the entire Midwest, would get us in return only about one-fourth the energy that the United States presently consumes in oil every year.[256] This is obviously a bad idea.

We might imagine avoiding competition with food by growing special energy plants with no food value on lands presently too poor to farm. That is a bad idea too, as it turns out, because there are no such plants. Food energy and wood energy are the same thing as far as a plant's concerned, so a woody or weedy plant won't produce well on marginal lands for the same reason a food plant won't. Thus we don't gain anything by planting, for example, tall prairie switchgrass, a perennial bioenergy favorite, as a fuel crop. It likes the same kind of land corn does, being native to the corn belt, and thus displaces corn when it grows. It also produces about the same amount of energy per unit of land that corn does—or perhaps slightly less because corn is an extraordinarily efficient plant.[257] Tall prairie switchgrass will indeed grow in the desert with lots of help from people, but then so will corn—as any farmer west of the Rockies will happily explain to you.

Accordingly, although the agricultural practices that supply food two hundred years from now are likely to be the same as today's (they haven't changed fundamentally in centuries), the practices that supply carbon for fuel are likely to be radically different. The reason is simply that if they aren't, they'll compete for resources with the food industry, a political fight they cannot possibly win.[258] We can expect some marginal contributions from conventional agricultural waste products such as sawdust or sewer gas because they're nuisances we have to dispose of anyway, and the temptation to exploit high fuel prices will be irresistible, but none of these things can scale up sufficiently to handle the stupendous masses of carbon required to keep the world economy humming. Given how thinly stretched the world's freshwater supplies presently are, it's difficult to imagine any version of this parallel energy agriculture system not based on saltwater cultivation. That narrows the field a bit, for there just aren't that many crops that will grow cheaply in saltwater.

The plants most likely to be exploited in the long term as an industrial carbon crop are saltwater microalgae.[259] This prediction is somewhat of a stretch, because there is no saltwater agriculture industry at the moment from which you can make crop comparisons. However, microalgae grow, and thus fix carbon from the atmosphere, faster than any other plants in the world, so they're very strong candidates.[260] Their near-term production cost per unit mass is probably between five and ten times coal's.[261] This makes algae agriculture badly uncompetitive as an industrial carbon source while coal is plentiful but a potentially acceptable one when coal is exhausted.[262] Other saltwater plants would have to beat this cost to win in the marketplace, and none of them can do so at the moment.

Microalgae also have an immense public relations advantage over other potential energy crops in being quintessentially green. They're not just natural things compatible with the environment; they're the fundamental food source for the ocean's entire ecosystem. These free-floating one-celled plants are visible at most ocean beaches as a green or greenish-blue tint of the light passing through the peaking surf. The powder blue color we find in open sea or in warm tropical waters indicates that the nutrients necessary for aggressive algae growth, chiefly nitrates and phosphates, are in short supply. But given a flow of nutrient-rich coastal runoff or a deep water upwelling, as we find on the Grand Banks of Newfoundland or Peru's Humboldt Current, the water becomes turbid and green and abounds with life. Filter feeders such as shellfish, corals, and larval zooplankton eat the algae and multiply with abandon. Fish eat the filter feeders—and each other—eventually becoming abundant. Higher predators such as birds, seals, and killer whales eat the fish and likewise prosper. Commercial fishing operations thrive. The central role that algae may eventually play in the human economy when the coal runs out is thus extremely appealing, in that it echos the role they already play in the sea.

Unfortunately, algae are difficult to farm.[263] The fundamental reason why is that they achieve their great fecundity by avoiding responsibility, not by being gifted. A conventional food plant like corn has stems to erect, roots to put down, leaves to unfurl, seeds to generate, and so forth, not to mention longer-term things such as waiting for spring. But algae do none of these things. They lead a short, brutal life of freedom rather than a long one of endless toil, and they just don't care about tomorrow. As a result, we have to do their chores for them if we want to raise them, and that costs money. The extra costs wind up being so great that no one can figure out how to farm algae profitably. That isn't so surprising, for pure algal biomass is a green substitute for coal, a resource that is extremely cheap at the moment and correspondingly hard to beat.[264] But the problem is worse than that, for algae become less productive than conventional agricultural plants, not more productive, once we discount the extra costs of raising them. Algae farming will probably not become a significant industry until both coal supplies and land resources for agriculture have run out.

Algae's monumental cost problems make it almost impossible to take present-day algae biofuel companies seriously.[265] We want to believe, but we find ourselves thinking instead about all those nutritious government subsidies. The most unforgettable of these subsidies is the high-priority U.S. congressional mandate that the Air Force secure a supply of green jet fuel at any price.[266] Exactly how much the Air Force is paying isn't public knowledge, but rumors are that it's about ten times the cost of jet fuel made from petroleum.[267] Not surprisingly, saltwater agriculture is a rather low priority with startups jockeying to supply this fuel, as is agriculture generally. For example, one of them is reported to be growing algae in plastic bags (made from petroleum) stacked in warehouses.[268] Another isn't engaging photosynthesis at all but is instead grazing its algae in the dark on cheap sugar, presumably obtained from conventional crops such as beets or cane.[269] Another eschews growing algae completely and plans instead to harvest the fish that eat them, no doubt with the objective of grinding the fish up and processing the happy brew into gasoline and diesel fuel.[270]

The absurdity of green biofuels that never see the sun is actually not funny, as it's symptomatic of attempts to contravene the laws of economics—presumably out of fear of what those laws may portend.[271] If the big energy companies won't make algae biofuels, so the reasoning goes, we must do it for them, thus heading off impending catastrophe when the oil runs out. But there is no need at the moment to create saltwater agriculture that spares land and water for food production. If there were, we would have such agriculture. There also is no need to generate as much biomass as we possibly can as a carbon resource. The world is awash with cheap carbon fuels. The world also has agricultural land to spare. It pays farmers not to plant the fields they already have.[272] It struggles to find buyers for subsidized butter and cheese in oversupply.[273]

The energy industry's sudden interest in algae might also be part of this absurdity, unfortunately. Green politics powerfully encourages "greenwash," the practice of associating yourself with green causes to look more environmentally friendly than you actually are. Although the investments that the oil majors are presently making in algae look technically legitimate, they might just be public relations expenditures.[274] We can't tell, for the amounts of money involved, though considerable, are smaller than the potential costs of taxation, regulation, and political vexations that might be visited upon them for not being sufficiently green.[275] Absent some truly unprecedented discovery or breakthrough, it will be hard not to smile knowingly whenever world-famous geneticists begin explaining their strategic algae oil partnerships that involve no farming until sometime way in the future, if ever.[276]

Ironically, algae are easy to farm theoretically because all we need do is overfertilize them. This induces a population runaway called a bloom, a kind of plant version of a nuclear explosion. The growth rates we get are blistering—in some cases four times the seasonal average of the best corn fields. These artificial blooms emulate natural ones caused by nutrient upwellings in all the world's oceans. Natural blooms are often hundreds of kilometers in size and can be seen from space by satellites.[277] Rates of biomass creation in them can approach traditional food crop values, even though the blooms aren't farmed.[278] Some of this blooming is actually important for healthy ocean ecology. A well-known example is the spectacular phytoplankton bloom that occurs as pack ice retreats in the antarctic spring. This bloom causes krill populations to explode, thereby enabling great blue whales and armies of penguins to stuff themselves in the southern oceans.[279] Blooms can also be detrimental, as they are, for example, when noxious red tides sweep in from the oceans and poison clams. They also cause dead zones, such as the famous one just off the coast of Louisiana, where zealously producing algae die, sink to the bottom, use up all the oxygen by decaying, and thus kill the fish.[280]

Algae's practical biomass production rate is probably about twice corn's.[281] Nobody knows the precise number because commercial algae farms don't yet exist, but this is a reasonable guess given what's known. The twofold advantage only applies to warm, sunny climates, however, such as Hawaii or the deserts of Southern California. Algae farmed in places where there are seasons stop growing in the winter and thus produce, as a yearly average, roughly the amount of biomass corn does. Corn, it turns out, has growth spurts too, and it's a tough competitor. Assuming that the weather is warm, so that we can grow the algae year round, the amount of pond area required to produce the energy equivalent of all the oil presently consumed in the United States is about twice the area presently planted in corn, or slightly more than the total area of Texas.[282]

The long-term cost of transport fuel made from algae when the coal runs out is likely to be about twice the present cost of transport fuel made from petroleum, not ten times more, as present-day estimates suggest. The reason is simply that costs will lower over time in response to competitive pressures and improving economies of scale. The sole exception will be farming costs, which will then become the main price-determining factor, just as they are today in the food industry. The farmer's cost to raise a given dry weight of biomass can be reasonably estimated at the cost of raising the same dry weight of corn silage. The cost of a given dry weight of corn silage, in turn, is about twice the present-day cost of the same weight of coal.[283] But the price of coal is pegged to the price of conventional transport fuel through the implicit threat, if the latter stays high for too long, of competition from Fischer-Tropsch synthetic fuels. Thus the long-term cost of algae-based fuel probably cannot exceed the cost of Fischer-Tropsch fuel made from coal, pro-rated by the price difference between farmed corn and mined coal. That lies somewhere between two and three times present-day fuel prices, depending on the relative importance of the hydrogen and energy content of the biomass.

The industrial processes by which people convert this agricultural biomass into transport fuels (and carbon-based products generally) will probably be brute force gasification followed by catalytic synthesis. It's conceivable that microbial processing will become cheaper than those methods and supplant them, thus reducing the amount of dirt and noise people have to put up with. But it's a long shot because carbon, not energy, will be the expensive commodity by that time. To be competitive microbes would have to both maximally exploit carbon and make molecules suitable for trucks and cars. That's a lot to ask of them.

If industrial chemistry does indeed prevail in the competition to render biomass into useful products, then ethanol probably won't be a major component of transport fuel, notwithstanding the enormous public investments being made in it now. Ethanol will continue to power wine, beer, and spirits, of course. The reason is not that ethanol is a bad fuel, which it's not, but simply that ethanol manufacture would throw away too much carbon.[284] Right now, ethanol's chief advantage over other gasoline substitutes is that we can make it from cheap corn and sugar cane surpluses with ancient biological brewing techniques known to everybody and in facilities that don't cost very much to build. However, this convenience comes at the price of discarding most of the carbon in the feedstock. Ethanol's economic advantage will thus weaken when carbon becomes expensive.

One reason biological processing tends to use carbon inefficiently is that the microbes doing the job for us don't work for free. We think of microorganisms as mindless little slaves that will do whatever we want, but actually they're scheming, selfish entrepreneurs who don't give a fig about helping us and only care about multiplying and filling the earth with their progeny. In fact, some of them are willing to eat us. If we can work with them at all, which normally we can't, we have to find them a job they like and pay the price they demand. That price is usually food. In conventional ethanol production, the yeast doing the fermenting want to be paid in sugar and will happily produce alcohol from this sugar as a metabolic waste product if we deny them oxygen. They do want a cut of the energy take of between 10 and 20 percent, but it's a reasonable price to pay.[285] However, they'll ignore all parts of the feedstock that aren't sugar.[286]

The other, more important reason for these carbon inefficiencies is that they aren't necessarily wasteful from the perspective of microbe economics, the currency of which is energy. Microbes live on energy, just as people live on money, and they make decisions about what to do based on how much energy profit they can gain. Living on their own in the wild, they prosper if they make good energy decisions, suffer if they make bad energy decisions, and die if they make really terrible energy decisions. They've been living with these cruel facts of life for three billion years, so they really know what works for them business-wise and what doesn't. Accordingly, tricking microbes into doing things they don't normally do in the wild usually comes at an energy cost. This cost doesn't matter to the manufacture of medicines, which is what most of biotechnology is about, for we can just feed the microbes extra sugar (made from corn) to cover the loss and then pass on the charge to patients. It matters critically to the manufacture of fuels, however, because we and the microbes are then after the same thing, namely energy. For example, genetically engineering yeast to make ethanol out of cellulose, the main component of woody plant material, would be quite counterproductive if the energy the yeast used to make ethanol this way exceeded the energy contained in the ethanol thus made. Doing so and then feeding the poor thing only cellulose would condemn it to cannibalize its own body to subsidize fuel generation for us, a practice that would soon dispatch it to the Great Fermentation Vat in the Sky. It might be that wild yeast don't eat cellulose because they're technologically challenged, but it also might be because they're smart.

Whether or not microbes can profitably render cellulose into fuel matters a great deal to the carbon sector of the future energy economy, as it turns out, because cellulose constitutes about one-third of all plant dry biomass.[287] Firewood, for example, is about 40 percent cellulose by weight, as is corn stover. Cotton fibers are pure cellulose. Even algae, the putative fuel plants of the future, are about 20 percent cellulose and hemicellulose.[288] Thus, if microbes can be coaxed into getting the energy out of cellulose, then pure biorefining will stand a much better chance of successfully challenging the traditional chemical industry.

However, cellulose is deliberately designed to be difficult to digest. In fact, it's one of the all-time great technical innovations of plants—a strong structural polymer made by chaining sugar molecules together that is, nonetheless, so difficult to convert back into sugar that it's nearly useless as food. Exactly why it's so useless is one of biology's best-kept technological secrets, but there can be no doubt that it's true. Trees continue to grow tall today, four hundred million years after plants invented wood.[289] Fungus and termites eat wood, but not very fast or easily, presumably because their energy profit in doing so is low. Cellulose certainly contains plenty of energy, as any campfire nicely demonstrates—indeed slightly more energy than the sugar from which it's made.[290] But no organisms, including the plants that made the cellulose in the first place, seem to be able to get this energy out easily. This intransigence is fortunate, in a way, because all our furniture, paper, house frames, cotton underwear, and so forth would instantly disintegrate if metabolizing cellulose were easy. However, it's also a serious impediment to making purely biological fuels.

Microbes digest cellulose by means of enzymes called cellulases. These are large, sophisticated proteins that cleave the cellulose back into the glucose from which it was originally polymerized. They use up no energy as they do so and return back to their original state after a given cleave is completed, ready to proceed with the next—ready, that is, until they break, which all sophisticated machines eventually do. This breakage is central to the energy calculus of cellulases, for these great lumbering molecules cost a lot of energy to synthesize. The original source of this energy is glucose, the very fuel their function liberates. Exactly how the glucose budgeting works—glucose liberated versus glucose expended to do the liberating—in organisms that digest cellulose isn't known exactly, but it must be touch and go because underwear-eating fungi have not taken over the world.

Cellulases are pervasive in modern life, as it turns out. They've been important commercially for over twenty years and presently comprise about 14 percent of the $1.1 billion industrial enzyme market.[291] They're chiefly used in the textile industry to make prefaded blue jeans, a process called biostoning. They're also added to laundry detergents to remove accumulated fuzz from cotton fibers, thereby freshening fabric colors. Commercial cellulases come from wood-digesting fungi, in particular one *Trichoderma reesei*, a native of New Guinea that came to everyone's attention during World War II through its obnoxious tendency to eat cartridge belts of U.S. soldiers in the South Pacific.[292] A mutated variety of this organism was subsequently discovered that produced twenty times more of the relevant enzyme, and that's the one cultured today.

Thus, digesting trash cellulose, such as corn husks and wood chips, with commercially available cellulases and then feeding the glucose thus generated to yeast to make ethanol fuel is technically feasible right now. This would, in principle, open up vast nonfood feedstocks for biofuel manufacture. The point of not competing with food is somewhat inaccurate, of course, because the glucose intermediary would be no different from conventional corn syrup. However, the genuinely serious difficulty with this idea is that the glucose cost of making cellulase is high. Exactly how high isn't public knowledge, for the manufacturing process, which involves cultivating fungi and harvesting the enzymes from them, is proprietary. But if the amount of glucose required to make the enzyme were large, as the marginal lifestyle of fungi suggests might be the case, then cellulosic ethanol would burden the world's food resources just as much as corn ethanol does now. The reason is that a given amount of ethanol would require, in addition to cellulose feedstock, lots of precious glucose (made from corn) fed in behind the scenes. Unfortunately, the cost of corn syrup is presently so low that quantifying the glucose budget of enzymatic ethanol is difficult. The best we can do is shop for cheap enzymes. That's why one becomes concerned when enzyme people not receiving cellulosic ethanol subsidies estimate the enzyme cost to make ethanol at about four times the market price of the ethanol.[293] An incisive total energy measurement or two could easily sweep away this concern, but such measurements are hard to find, for whatever reason.[294]

There are, of course, nonbiological methods for converting cellulose to ethanol that might be used in the future.[295] These, however, are not new and would also have to hold their own economically against industrial processes that convert biomass to other fuels, notably conventional gasoline. Ironically, one of them is a conventional syngas process: heating the biomass in the presence of steam to make carbon monoxide and hydrogen, followed by fuel synthesis from the gas mix. It differs from conventional synthetic fuel only in the final step, which produces ethanol instead of gasoline. Describing this fuel as cellulosic ethanol, which many people do nowadays, is amusing. It's reminiscent of the famous Mark Twain story in which a man named Dunlap insists he is French because his actual name is D'un Lap.[296] Another method is the Bergius process, in which cellulose is hydrolyzed to glucose by digestion in acid. That works quite well, but it leaves us with a noxious mixture of acid and glucose that we have to separate. This separation is expensive, as are the costs of purchasing, recycling, and disposing of the acid, an environmental pollutant. Moreover, if we're going to subject the cellulose to this kind of abuse, we might as well just go all the way and render it to syngas.

Ironically, one of the main criticisms of ethanol as fuel—that it requires slightly more energy to make than it delivers—may very well work in ethanol's favor when coal becomes scarce. The reason is that inversion of electricity and fuel prices will make it economically sensible to burn electricity to make transport fuel. We wouldn't do that today, of course, because electricity is the most expensive kind of energy. The chief villain in ethanol's energy shortfall turns out to be yeast's refusal to work when the alcohol concentration in its broth is larger than about one-fifth the value we need to make fuel.[297] Yeast plants are not likely to be coaxed to work at much higher concentrations through clever genetic engineering, for they stop mainly because alcohol is poisonous. That's why we use alcohol to sterilize medical equipment. Their decision to stop, in turn, requires us to further concentrate the ethanol by distillation, a process that uses large amounts of heat.[298] The energy equivalent of this heat is about half the energy the ethanol delivers as fuel.[299] If we get this heat from coal, as manufacturers typically do for price reasons, then about half the ethanol's energy content effectively comes from coal. Thus, the ethanol is not really all that green. But we can perfectly well supply this heat from electricity instead of coal. If we do, about half the ethanol's energy content now comes from electricity. This is a charming and potentially important biofuel concept, for no living thing has yet mastered the skill of turning electricity directly into gasoline, at least at levels relevant to industry.[300]

But even if there's profit in putting electrical energy into biofuel instead of taking it out, ethanol will still have to compete with conventional syngas made by heating biomass electrically instead of raising its temperature through combustion.[301] This approach has economics similar to ethanol's, and it has the advantage of producing conventional gasoline and diesel fuel instead of alcohol. It also doesn't compete with food supplies—unless, of course, one is foolish enough to use food as feedstock.

In light of biofuel's present competitive weaknesses, people likely will be searching between now and the onset of coal exhaustion for improved biological processes that produce energy from things we can't eat and don't want. There is likely, for example, to be continued interest in termites, which seem to be better than fungus at attacking wood.[302] The reason they do better is probably just that they chew the wood up, thereby speeding digestion by the cellulases generated by microbes in their guts, but it's always possible that a particular microbe in there beats New Guinea fungus when it comes to digesting cellulose. There will no doubt be proposals to turn termites into diesel fuel, along the same lines of doing it with fish. Depending on competition for food supplies, someone is bound to get the bright idea of doing it with cows because that would cut out the middle man and supply serious fuel instead of merely manure vapors to power homes. There will continue to be fresh press releases about genetic engineering breakthroughs in which bacteria turn wood chips and grass clippings (but really glucose) into petroleum.[303]

Meanwhile, the real microbes, unfazed by any of this, will continue leading their lives as they have for billions of years, opportunistically seeking out energy profits for themselves wherever such profits can be found. That will include not only beer vats and ungulate rumens but also fuel tanks, oil wells, and the La Brea Tar Pits, because microbes, unlike some of their more confused human cousins, know where the energy is, and that there can be no free lunch.[304]

CHAPTER 8

Trash
Ash

It's a lovely morning two centuries from now. Dawning light floods over the municipal landfill, painting it gold. Reflections flash here and there from shards of broken glass and metal scattered about among grapefruit rinds, broken lawn chairs, and cardboard. All is quiet. A light breeze comes up and blows colorful bits of paper about in a casual way, like leaves on a forest floor. Skip loaders stand by cold and silent, ready to power up and meet the parade of trucks that is even now making its morning rounds, emptying neighborhood trash cans. Seagulls have begun wheeling about, occasionally swooping down for some goodie—perhaps a piece of spaghetti, a bit of pastrami sandwich in butcher paper, or a carton of supersize fries prematurely tossed. Behind them rise low hillocks of newsprint, glossy magazine paper, and rags, decorated here and there with broken furniture legs, mops, and broom handles, all casting long shadows in the morning light. Lots of plastic is in there too—juice containers, soda containers, food wrappers, empty pill bottles from the medicine chest, and what have you—beaded with dew that won't sink in.

Garbage dumps—those ancient and venerable institutions of mankind—are unlikely ever to change much.[305] They're the end of the human value chain, the place where things go when their previous owners don't want them anymore and can't sell them to anyone else. There's more trash now than there was in the past because there are more people and because consumer items are cheaper, but the underlying economics and psychology of trash are the same as they've always been. Nobody cares about dumps or their contents, by definition. Dumps are where we put things we don't care about. There will always be a need for such places because all healthy human beings generate a steady stream of things they no longer care about. It's part of life. Even the most ardent environmentalist will, from time to time, put things quietly in the dump. But whereas trash generation is natural, it isn't pretty. Trash is fundamentally an expense we have to bear, a problem to pass onto someone else and forget as soon as we can.

However, trash contains a great deal of carbon. That means it will become less problematic a century or two from now when carbon becomes dear. Transport fuel companies, in particular—which by then will be forced to purchase carbon from farmers—will not be amused to see the otherwise free carbon in trash buried in landfills or burned. Whether they exploit the trash for fuel feedstock or allow it be disposed of by traditional means depends on processing costs and government regulation and subsidy formulas, all of which are difficult to predict. But trash carbon will, at the very least, impose a price constraint on the amount farmers can charge for their biomass and thus a constraint on the price of fuel. Consequently, household trash is destined to play a significant role in the energy economy, if only as a price regulator.

The amount of carbon in trash is immense. In the United States alone it amounts to eighty-eight million tons per year, or about one-tenth of the carbon in the coal (or oil) that the United States burns in one year.[306] To put this number in perspective, we could remove all the carbon dioxide dumped into the air by industrial civilization thus far plus all the amount destined to go into the air through future fossil-fuel burning profligacy simply by passing laws forbidding both recycling and the burning of trash for any purpose and requiring refuse companies to bury the entirety of their collections in landfills. Such laws would have a significant effect on the atmosphere in about a millennium and would remove every atom of excess carbon dioxide in about five.[307]

Nonetheless, this amount of carbon is wholly inadequate for supplying the world's energy. The 250 million tons of municipal garbage that a big industrial country like the United States generates every year may sound like a lot, but it's insignificant compared to the two *billion* tons of coal and oil that the United States consumes annually.[308] If dried, trash contains roughly the same energy per unit weight that coal does. Thus, if we burned all U.S. trash to generate electric power—which would extract its energy optimally—the total would add up, like the carbon content, to about 17 percent of the energy the

United States consumes in coal or 4 percent of the energy it consumes in all fossil fuels combined.[309]

Thus trash's potential importance to the energy future is not as an energy source but as a carbon source. This distinction is important, for burning trash to make electricity, which has serious profit problems today, will have even greater problems if electricity from noncarbon sources becomes cheap and abundant.[310] Generating electricity from trash under such circumstances would be like transmuting gold into lead—technically impressive but not very clever. Carbon for making transport fuel, not the energy the transport fuel contains, would be the thing in short supply and thus the thing of value.

Trash's shortcomings as an energy source illustrate rather starkly how difficult it would be to supply all the world's energy, as opposed to transport fuel carbon, from biological sources. The biggest component of trash is paper, the bulk of which is made from wood pulp. Even if we diverted all the wood consumed in the United States to energy production, not just the third of it presently rendered to pulp—thereby stopping all housing construction and repair in its tracks—and even if we could keep up this burning pace in perpetuity, we'd still be supplying only a quarter of the energy the United States presently obtains from coal.[311] Supplementing coal furnaces with wood chips, as is sometimes required by law today, is thus a nice symbolic gesture but not a very practical way to prepare for the end of coal.[312]

Trash's shortcomings also reveal why we can't supply the world's energy with other combustible components of trash, such as plastic. The amount of plastic in trash is very large—about a third the amount of paper and wood—and plastic contains more energy per unit weight than paper and wood do. Nonetheless, the total energy in plastic is less than the total energy in paper and wood, so it's therefore also inadequate. The reason for its inadequacy, however, is not that agriculture is strained but that plastic is made from oil.[313] It's a shockingly small fraction of the oil, too—about 5 percent. The rest goes primarily for transport fuel. Thus, even if we processed all plastic trash into transport fuel, we'd reduce the world's oil consumption only by about 5 percent. For the same reason, we can't save significant amounts of energy by recycling plastic. Even if we refused to burn or reprocess any plastic but instead used all of it over and over again forever, thus eliminating all the plastic in trash and also the entire petrochemical industry (because it would no longer be needed), we'd only reduce the world's oil consumption by about 5 percent.

The method most likely to prevail in extracting trash's carbon for transport fuel is heating the trash in the presence of steam to make synthesis gas. The argument that this will occur is the same as for biomass, to which trash is chemically similar, except it's a bit clearer in this case because trash will never be manipulated biologically. Nobody is going to find a bug that turns all the awful stuff in New York City garbage into gasoline. They'd be lucky to find a bug that could survive in there without getting mugged by the other bugs. But rendering the trash into synthesis gas is relatively straightforward. This is demonstrated by, among other things, the present-day proliferation of trash gasification startup companies.[314] Nearly all of these companies are proposing to burn the gas they produce in turbines to make electricity, not to make gasoline or diesel fuel, but this is mainly an issue of cost. At the moment transport fuel energy sells for about as much as electric energy, but it requires a much larger capital outlay to deliver and thus yields much smaller profits.[315] Going further to make conventional fuel from the gas requires only borrowing technology used to make synthetic transport fuel from coal. This technology is generic and can be used interchangeably among plants using coal, trash, and biomass as feedstocks. Thus the existence of these startups is one of the more persuasive indicators that the full-blown biofuel industry, when it finally arrives two centuries from now, will be built primarily on synthesis gas.

However, nobody is likely to make meaningful amounts of synthetic fuel out of trash until oil becomes scarce. The reason is simply that price competition between the petroleum and coal industries will continue to block profitability. Coal, in contrast to trash, is available in stupendous quantities and could supply the world's entire transport fuel needs, not just some of them, if market conditions were right. Oil companies, therefore, have a powerful incentive to keep their prices low enough to prevent market penetration by synthetic fuel made from coal. That, in turn, prevents market penetration by synthetic fuel from trash, which has similar processing costs. An analogous thing happens with synthesis gas itself, except that in this case, it's the natural gas industry that must keep prices low to make sure coal is kept at bay.

But all of these blockages will give way as fossil fuel depletes. As oil and natural gas become scarce, prices of transport fuels will rise until they reach those of synthetic fuels manufactured from either coal or biomass. When this happens, fuel from trash will very likely become profitable at the same time. It's difficult to say for sure because trash will still be regulated, but trash feedstock has the fundamental advantage over all of its competitors in being worthless to its previous owners. It's therefore reasonable to guess that fuel made from trash will be slightly cheaper than fuel made from coal. If it is, then the trash disposal problem as we know it today will totally disappear. Trash will just be too valuable to throw into landfills or even to leave lying around. The increasing scarcity of fossil fuel will, in other words, have become the world's most effective deposit law.

Garbage is important not only as a potential future energy price regulator but also as a bellwether of environmental troubles sure to arise in the future as the energy industry restructures. Trash burning is more visible than, say, coal burning, which damages the environment much more but works its mischief out of sight, making its effects easier to ignore.[316] Coal pollution is something we'll always think about tomorrow. But everybody knows what a nuisance trash is; how getting rid of trash costs a fortune; and how burning trash produces smells, smoke, and poisons that we want released, if at all, somewhere else. So concerned citizens, sure in the knowledge that they know how to solve this terrible problem (with other people's money), communicate their wishes to the authorities, who respond in Alice-in-Wonderland fashion because that's all they can do when asked for impossible things: *Separating trash for recycling is crucial because it reduces consumption of energy, conserves natural resources, and relieves stress on the environment, except that we're not recycling lots of things this year because we can't find buyers.* Or, *trash incineration companies should make synthesis gas instead of just burning the trash directly to make electricity, because incineration is bad, and we don't know how much synthesis gas manufacture pollutes, and anyway we've banned incineration and allocated money to truck our trash to landfills in distant states.* Contemporary experience with trash makes it easy to imagine how biomass power and fuel generation might also be environmentally sensitive and how deciding who will suffer the plants in their backyards and who will pay for health and quality of life sacrifices might conflict with deciding how much energy gets generated, who gets it, and what its price will be.

The ultimate recycling from the earth's point of view is, of course, burning. That includes both incineration of trash and the combustion of fuel made from trash in engines. The carbon dioxide and water that result from burning, if it's complete, are the feedstocks that plants use in the manufacture of their bodies. All of the carbon in the paper part of trash, for example, was in the air a few hundred years ago, so returning it there after we're done with the paper makes perfect environmental sense. We're just completing a natural cycle that we usurped by cutting the tree down before it could die of old age and decay (or succumb to a forest fire). The carbon in cotton rags and leather in the trash likewise came from the air, except even more recently, so it makes even more sense to return it to the air when we're done. Plastic is more problematic because it comes from oil, which was not in the air recently, although it probably was in the air several hundred million years ago. But plastic buried in the ground is a pollutant, whereas carbon in the air isn't, so the environmentally responsible thing to do is to burn plastic when its useful life is over. The effect of doing so on carbon dioxide levels in the air is marginal because we're burning most of the petroleum anyway, and we can later return this carbon to the ground, if that's our intention, as biomass.

But recycling carbon atoms by burning to completion isn't profitable. This fact is central to trash economics, for charging people to take things off their hands only works if we can rid ourselves of those things at low cost. That's why most trash even today gets thrown into landfills.[317] We could dispose of it other ways, but the customers (voters) don't want to pay more in dumping fees than they absolutely must.[318] Moreover, they're resourceful. If we attempt to charge more than they think is fair, they'll just dump their loads on some lonely country road late at night when no one is looking. When landfill space in a town becomes scarce and thus expensive, people can advance to more sophisticated disposal strategies, but often they can only afford the next cheapest one, which is sloppy burning.[319] In its most primitive form sloppy burning entails drying the stuff in the sun for a few weeks, heaping it up in a big pile, and lighting it off with a match. The bonfire can be quite glorious, but it smells bad and sends up lots of greasy smoke and cinders into the air. The industrial version of sloppy burning involves a building outfitted with a big furnace, but what goes into the air is functionally the same. Most municipal incinerators do a better job than a bonfire, but none of them is perfect, and incinerator operators are always fighting tradeoffs between emissions and cost. Synthesis gas processing, though conceptually more sophisticated, also releases pollutants and also has cost tradeoffs.

The poisons released by the incomplete burning of trash are legend. The worst of them by far is ordinary soot, which contains a witch's brew of complex hydrocarbons, most of which irritate lungs, many of which asphyxiate us in even small concentrations, and some of which cause cancer. These substances are not unique to trash but are also produced in great abundance in natural burning events such as wildfires and also in everyday things such as the fumes of a candle we've just snuffed out or a smoky diesel truck exhaust. Fortunately, they're all vulnerable to attack by sunlight and the oxygen in the earth's atmosphere and probably by many microorganisms as well, so they soon degrade to harmless carbon dioxide and water and disappear. That's why, for example, the blackened remains of terrible grass fires become difficult to detect after a few years. But trash also contains things that ordinary biomass doesn't, such as chlorine from vinyl plastic as well as mercury, cadmium, and lead from exhausted batteries, and these increase the dangers beyond those of conventional smoke.[320] The exact size of the increase is controversial, but everyone agrees that we don't want to be immediately downwind of a low-budget incinerator.

Thus full recycling of trash carbon into the air is destined to conflict perpetually with public safety laws. In theory we can burn trash so completely that it doesn't hurt anybody, but in practice costs constrain how clean the exhaust can be.[321] Natural environmental processes will eventually render escaped pollutants harmless, but not necessarily quickly, and anyway this is little comfort to individuals living near the plants. That's why nearly all governments regulate incinerators. We know from notorious examples such as Los Angeles smog what will happen if they don't. Most people will choose low cost over clean air if the money required to clean up the air is their own. This aspect of human nature is very primal and thus unlikely to change even as fossil fuel supplies tighten. It may even worsen in response to the tightening, because transportation is a significant portion of trash disposal cost.[322] Unfortunately, however, there is no natural measure for how dangerous residual industrial pollutants are to human health, and thus no natural procedure for crafting the laws regulating them. The process instead involves complex compromises among various parties—particularly as to who will be paying for health benefits for whom—that are inherently confrontational. This can be expected to remain the case long after the fossil fuel crisis has passed.

However, the conflict with health laws is likely to shift in favor of industrial processing when oil and natural gas begin to fail. The reason is simply that making synthetic fuel out of trash will have become profitable, whereas landfilling, its main competitor, won't. The municipalities write their own regulations, and they'll find it irresistible to slant them slightly in favor of lower taxes. The air is unlikely to become dirty as a result, for even conventional incineration is thorough if the incinerator is modern, but the air will be less clean.

The health issue of trash likely to cause the most trouble as fossil fuel depletes is thus not air pollution but ash pollution. In contrast to impurities such as chlorine, which find their way into the air, cause lots of trouble there (i.e., as dioxin), and then degrade to benign substances such as table salt, the metals in trash are intractable. Incineration doesn't burn them away. Oxygen, sunlight, and microbial action don't render them into harmless carbon dioxide and water. If anything, environmental action makes metals more troublesome by corroding them. At the moment, the heavy metal content of trash is relevant mainly to distant landfills, so we ignore it for the same reason we ignore coal-burning headaches. But this will change when trash gets systematically exploited for carbon content. Extraction of synthesis gas from trash, like in-cineration, concentrates metals in the ash left behind.[323] This ash is visible, abundant, and toxic, and that makes it newsworthy.

Ironically, ash's surly reputation is largely undeserved.[324] Burning trash doesn't fundamentally alter its chemistry, nor does it change the amount of metal involved. Ash is, in this sense, exactly as dangerous as the trash from which it came—no more, no less. The main effect of burning is to concentrate the metal, which makes it a more aggressive pollutant by speeding up natural leaching processes. When rain falls on landfills, chemical action dissolves the metals in them as ions. The ion-laced water then percolates downward, even-tually finding its way into the groundwater and, from there, to wells and springs. However, the extent of the resulting danger to drinking water is dif-ficult to assess on purely theoretical grounds because metal ions react chem-ically in complex ways with the rocks and soils through which they percolate.[325] In this they differ fundamentally from the organic solvents that sometimes find their way into groundwater. Water hardness, an annoying fea-ture we find commonly in wells around the world, comes from metal ions leached out of the earth, not from metal ions loosed on the world by people.[326] This same process working in reverse can make dangerous leachate ions dis-appear into the rocks. The danger to drinking water is also difficult to assess experimentally, for it involves digging test wells and routinely monitoring the water in them, all of which costs money. But even after accounting for these financing difficulties, we find that well-documented examples of drinking water poisoning by metals from landfills are unsettlingly scarce.[327] Also, the potential health threat that ions pose depends on the metal in question. Iron, copper, and zinc, for example, though harmful in large concentrations, are es-sential to good health in small ones, and are even ingredients of over-the-counter mineral supplements people buy in stores.[328] Conversely, lead, mercury, and cadmium are cumulative systemic poisons harmful in even very small concentrations if exposure to them persists.[329]

Thus, the ash problem is actually a matter of costs, just as the incinerator exhaust problem is. If money were no object, we could easily eliminate the ash threat completely by sealing the ash up in large concrete boxes. Ash con-stitutes about 10 percent of trash's mass, so the total amount of ash potentially generated by the entire United States in one year is about twenty-five million tons, or nine cubes the length of one football field on each side.[330] Construct-ing boxes of this size would be perfectly feasible technically, as would trucking the nation's ash to the boxes, dumping it in, and sealing the boxes when they're full. Big boxes would also have the advantage of enabling us to leach the ash on our own terms, say, with internal sprinklers, and to reprocess leachate to recover its valuable metals. Money is an object, of course, so we have to make cost compromises. The preferred method for disposing of ash at the moment, other than simply dumping it, is a low-budget version of the box in which we mix the ash into concrete and asphalt and use it to pave roads and reinforce buildings.[331]

Failing to deal with the ash is probably not an option, unfortunately. Suppose, for example, that we decided to dump the ash into the ocean. We wouldn't do this literally, of course, unless we were very thoughtless, for the metals in ash are even more poisonous to life in the ocean than they are to life on land, on account of being mobile in water.[332] We'd also be contravening international treaties and U.S. law.[333] However, the earth itself is the original source of all these metal ions, and it has many strategies for rebalancing local excesses of them. Mineral springs that issue brightly colored (and poisonous) water have existed since the world began, but they have not destroyed life.[334] The earth incorporates some of this flow of metals back into its rocks as ores. It also traps some in its soils, this being the reservoir that plants and, thus, the animals that eat them, tap to obtain the trace amounts of metals they need to live. But the rest of it washes out to sea. Indeed, the sea is a veritable cesspool of ions washed out of the rocks over the span of geologic time.[335] That's why it's salty. Why some ions are more numerous in seawater than others has to do with rates of erosion and precipitation of minerals on the ocean bottom, but ultimately it's unimportant. The plants and animals living in the sea are adapted to the metal concentrations we find there, so they can endure small amounts of even very poisonous metals without harm. Thus, we might leach the metals out of our ash intentionally, dilute them to concentrations below those found naturally in seawater, and pump the diluted leachate out to sea. Regrettably, the flow of water required to perform this task would be about fifty Amazons, even for the United States alone.[336]

Accordingly, governments will be increasingly moved to reduce the amount of metal delivered to trash streams, especially the exotic or poisonous ones. This has already happened, to some extent, with the passage of laws restricting the sale of consumer products containing mercury, forbidding the conventional disposal of batteries, and requiring the recycling of electronics.[337] These regulatory steps are just a foretaste of what's likely to happen as fossil fuels deplete, however, for although such laws are now mainly a matter of public health, they'll become a matter of money as well when carbon becomes valuable enough to extract from trash for profit. Everybody will be generating ash, and they'll find it easier to locate a final resting place for this ash if it isn't laced with problematic ions.

Unfortunately, this imperative to remove metal from trash is fundamentally incompatible with battery proliferation. No matter how solemnly battery manufacturers promise to buy back their old products and recycle them, and no matter how steadfastly they honor these promises (until they go bankrupt), some of the batteries, or parts of them, will always wind up in the trash.[338] Every one of these wayward batteries will contribute exotic metal ions to the trash stream because, as it turns out, exotic metal is the key working part of a battery. Not only must all batteries contain such metals, but they must also contain large amounts of them, for the battery's energy is stored in the metal itself. This problem is central to what batteries are and how they work, so it can't be evaded. Present-day searches for ever-better batteries are, as a practical matter, searches for ever-more exotic metals to put in them.

The imperative to remove metal from trash will thus interfere with present-day plans to create a great world fleet of electric cars.[339] The sheer mass of batteries in such a fleet would simply be too vulnerable to environmental attack. Even if these batteries lasted as long as cars do now, so that we never had to replace them until the car died, and even if we lost only 1 percent of the pesky things in the recycling step, which would be a very tall order indeed, their mass contribution to the trash stream would exceed that of dry cells entering the stream now.[340]

Political opposition to batteries and thus to battery-powered cars is likely to strengthen in the future because it's linked to fears for the earth. These fears aren't necessarily scientific, but whether they are scientific or not is irrelevant. Mercury is obviously a very poisonous substance that we don't want around in great quantities. Zinc and manganese, however, the main metallic components of disposable dry cells, aren't that poisonous, and so could logically be allowed in the trash at some level without harming water supplies. However, such logic is completely unacceptable to many people. The issue for them is not how hazardous these metals are now but how hazardous they'll turn out to be later after scientific research has exposed previously overlooked (or intentionally hidden) dangers—and, worse, after the environment has been irreparably compromised. The only certain way to protect oneself and one's progeny from these unknown future menaces is to ban substances from the environment that aren't already there in significant quantities. That means no batteries.

Battery proponents are thus likely to encounter future objection to things that seem perfectly innocuous to them now on grounds they find incomprehensible. For example, lithium, the working metal of the world's lightest batteries, is bound to become an environmental issue eventually, even though it isn't that poisonous, simply because it doesn't occur naturally in the environment. Very remote places such as the salt flats of Bolivia, where most of the world's lithium presently resides, don't count.[341] Lithium is also extraordinarily soluble in water, so it's likely to make its way quickly to water supplies, there to be easily detected by mass spectrometry on account of being exotic. It also affects the brain. Lithium salts are commonly prescribed to treat bipolar disorder.[342] Lithium ions will kill a person who ingests too many of them— but then so will a comparable number of potassium ions, which are chemically similar but also present in the body in great abundance and essential to life.[343] The workhorse of hybrid cars at the moment, the nickel metal hydride battery, is likely to encounter similar problems because of its rare earth metal content. The battery pack of a Toyota Prius, for example, which is not especially large by electric car standards, contains about twice as much metallic lanthanum as a conventional car battery contains lead.[344] Lanthanum is nowhere near as toxic as lead, but it's more toxic than zinc and, more to the point, more alien than either.[345]

Thus trash—the stuff nobody cares about, the as-yet unsolved problem that vexes mayors, eats tax money like no tomorrow, and blights the countryside—is likely to constrain the energy future in two important ways. One is to limit the price of transport fuel by constituting a large source of cheap carbon. The other is to disadvantage electric cars by causing their batteries to be declared toxic waste, whether or not they actually are, and banned. Underneath both is the reality that waste generation is a fundamental aspect of life, and the only kind of waste not poisonous to the environment is carbon dioxide.

CHAPTER 9

Viva
Las Vegas

It's a long drive from Los Angeles to Las Vegas. It isn't the time that gets you, for it's only four hours if you're lucky with city traffic. It's the endless dreariness of the desert. You drop down from Cahon Pass, full of hope that you'll soon see the distant skyline of Sin City welcoming you back to win again, but the unhappy reality is that you're not even to Barstow yet, and you have three more long hours to go. The interstate will carry you swiftly, but not swiftly enough, across a chain of vast and desolate valleys peppered with sage and tumbleweed but otherwise lifeless. Forbidding eroded mountains will rise up in the distance, and you must cross these now and then through hardscrabble passes, sometimes stony and sometimes not, each promising hope of life beyond the next rise but delivering none. There's nothing on the other side but another desolate valley identical to the one you just left. A lonely gas station punctuates the monotony now and then, perhaps accompanied by a fast food restaurant. This just adds to the eeriness, for they seem comfortably familiar until you remember where they are. You zoom by in air-conditioned comfort, but the windshield is hot to the touch, a reminder of how awful it is outside and what terrible things will befall you if your engine fails. The signs to places with colorful names like Devil's Playground and the Funeral Mountains make you smile because they're apt. And when you finally come over the last low crest and see Vegas spread out down below in the distance, you'll feel immensely relieved because you've made it to a place separated from the rest of civilization by a little stretch of hell.

What you may not realize, however, is that the stretch of hell you just traversed is the Saudi Arabia of solar energy. It doesn't look like the Saudi Arabia of anything except dirt, but looks are deceiving. The same conditions that make this desert so oppressive for you—the dryness, the lack of cloud cover, the absence of trees, the copious sunshine—also make it optimal for siting solar power plants. Other great deserts of the world, such as the Atacama, the Sahara, and the Arabian (in real Saudi Arabia), are equally well endowed with sunlight, but the Mojave is the only one also endowed with lots of power lines and close proximity to large, affluent metropolitan areas. Those things are just as important as sunshine when one is building power plants. Accordingly, the first great world city likely to become solar powered when fossil fuels begin to flag (if any do) is not Cairo or Karachi but rather Los Angeles.[346]

No cities of the world may become solar powered, of course. This kind of energy is notoriously expensive, and whether or not people elect to use it will depend on price and future events that are difficult for us to imagine today.[347] It's conceivable, for example, that people would choose to build nuclear power plants instead of solar plants in order to save money. This isn't very likely now, but things might be different when everyone has to get serious about costs.

However, solar power has a special place among potential future energy sources, being extraordinarily abundant. On a hot, sunny day at noon the sun sends down roughly one kilowatt (the power of a toaster) for every square meter of ground.[348] The sum of this power falling on the Sahara alone, after averaging over one day-night cycle, is three million billion watts, or two hundred times greater than the present energy needs of civilization.[349] Even if we reduce this number by a factor of thirty to account for the 3 percent efficiency of the most cost-effective (i.e., cheap) solar plants, we still have six times the energy needed to power the world.[350] If we do the same calculation with the western deserts of the United States, we get fifty times the present total energy needs of the United States—which reduces to twice those needs if we discount for the cheap plants.[351] Getting at this abundance would be difficult and expensive, but it wouldn't be impossible, and more to the point, it would be well within the means of future generations in trouble, as it were, over failed attempts to get by with something cheaper. There is also no possibility that civilization will ever outgrow this particular supply. The amounts of energy coming from the sun are simply too gargantuan. The lights of our cities may seem impressive when viewed from space at night, but their total power is only one-millionth of the total power of sunlight pouring down on earth's daytime side.[352]

Wind energy should properly be categorized as solar energy, even though it commonly isn't today. Exactly how and where the wind blows depends on mountain range locations, ocean widths, temperature history, earth's rotation, and other such details, but the reason the wind blows at all, and the reason it contains energy for us to extract, is that the sun's heat makes it do so. If the sun stopped shining, so that no heat flowed onto the earth or back into outer space, the earth's temperature would equilibrate, and all the winds would cease. In effect, the wind blows because the earth is a giant heat engine not so different from the one under the hood of a car, except that the working fluid is the entire atmosphere rather than some air gulped into a few pitiful cylinders. The machines that extract energy from the heat beating down on deserts are engines too, of course, just with different design details. Thus, solar energy and wind energy are really just two technologies for extracting energy from the same source—the sun—one with natural engines and the other with man-made ones.

The wind kind of solar energy has the obvious advantage over the hot desert kind that it requires no hot desert. The wind blows in all parts of the world, including exceedingly dark and cold parts that are anything but tropical.[353] Indeed, everyday experience tells us that winds in temperate latitudes are often greater in winter on account of the storminess.[354] This means that anyone can acquire energy from the wind—and also that the energy they acquire this way is secure. Countries lacking hot deserts are understandably leery of becoming dependent on hot desert countries for their energy.[355] Wind can also be harvested with simple turbine technology that's about half as expensive per given amount of power as a comparable solar facility.[356] It is thus not surprising that wind energy is between ten and one hundred times more widely deployed than solar-mirror and photovoltaic facilities, depending on how one counts.[357]

Unfortunately, all forms of solar energy, but especially the wind, are capricious. After all, they amount to harnessing the weather. This not a small problem but an enormous one. Energy is most valuable when it is dear, and people simply won't tolerate energy that blacks out at random moments. Exactly what fraction of the time—and for what reasons—the world's wind farms presently stand idle is politically sensitive, but the amount of energy they actually produce averages about 25 percent of their full capacity very generally.[358] Right now it's possible to paper over this problem by having natural gas peaking plants standing by ready to power up whenever the wind stops blowing— as it does from time to time.[359] However, this strategy won't work when the amounts of wind or solar power become large, because the correspondingly large backup generating capacity then will be too expensive. The people capitalizing this capacity don't like getting stuck with interest payments on machinery and gas tanks that generate no revenue on account of being idle most of the time, and they'll eventually just stop investing.[360] Of course, the strategy will have much bigger troubles than a few unhappy executives when there is no more natural gas left.

Accordingly, energy from the wind and sun can't become genuinely competitive until facilities capable of storing large amounts of the energy they harvest for later use are built. It's not completely clear that adequate numbers of such facilities will ever be built, for the costs involved are potentially very large, both in capital and in environmental compromises. However, there's a good chance that they will because the people of the future won't be facing a choice between storage and no storage but rather one between storage and the plutonium economy.

The most important of the many alternatives to covering the globe with hydroelectric lakes is storing energy as heat. In principle, we can store the kinds of energy we need with batteries, and we have already done so on a small scale.[361] However, doing it on a big scale isn't practical. For example, if the entire world's estimated supply of lithium, about thirty million tons, were made into lithium-ion batteries, it would store the present-day energy flow of the world for one and one-half days.[362] The entire world's estimated supply of lead, slightly over one billion tons, would do slightly better if made into lead-acid batteries, about two and one-half days. It's also possible to store the requisite amounts of energy by electrolyzing water and putting the hydrogen thus generated into tanks or underground caverns.[363] The capital cost of the machinery for doing this would be extremely high, the electrodes would need corrosion maintenance, and the masses of unwieldy hydrogen cyclically created and destroyed would be comparable to the mass of crude oil the world currently consumes. Thermal storage, by contrast, has none of these problems. It doesn't require any special materials because all substances have roughly the same heat capacity per atom at high temperatures. It doesn't have any corrosion issues because no electric currents flow through the storage media. It isn't a headache to handle the way hydrogen is because the important materials are all inert liquids and solids that don't explode. Also, it has been deployed in eight small commercial solar energy plants (and seven more in development or under construction), so it's market tested.[364] Except for long-term operating and maintenance expenses, which have not yet been fully field-tested (this takes forty years), its costs are known to be between one and three times those of pumped storage, depending on how one measures.[365]

Thermal storage has an interesting history. It was originally proposed as an alternative to hydropumping for leveling loads on nuclear reactors.[366] The idea was to run the reactor full tilt all day, even at night when demand was slack, and bank any extra energy that it made by heating up some large body—for example, the way one would heat a brick for later foot-warming purposes. Then, when demand exceeded the reactor's capacity during the day, the extra heat needed for powering the turbines could be extracted back out of the body. It wouldn't matter what the body was made of—rocks would do. However, there was a need for a fluid medium to get the heat in and out quickly, and this medium couldn't be water because of pressurized water's extreme explosion danger. This led eventually to the idea of using molten salts, cheap substances with fluidity in the right temperature range, low vapor pressures, and no tendency to dissolve or leach steels. Eventually it became clear that there was no need for the rocks at all. They were just nuisances that would crumble under thermal cycling and cause trouble in the machinery. The molten transfer fluid itself could be the heat storage body. It would then get pumped from place to place (while red hot) for heating and cooling purposes and stored in big insulated tanks. Although this technology failed to prove cost-effective for nuclear reactors, it was later picked up as a means of leveling loads on solar energy plants. The first field prototype was the Solar Two retrofit of the famous Solar One plant in Barstow.[367] The first commercial plant, Andasol-1, was deployed in 2008 in southern Spain.[368] Several other plants conceptually identical to Andasol-1 have since come on line.

The construction of Andasol-1 may have been a seminal event in the history of alternate energy—which is to say, of energy generally—although it's a bit early to tell for sure. The reason is that it provided proof of principle for the last technical element required for powering Europe with electricity from the Sahara. It also gave this proposition a precise price tag. It was already clear that the Sahara had the requisite energy resources and that underwater cabling and distribution technologies were up to the task. But any such plan without storage would have been nonsense because it would have required the lights of Europe to go out when night fell. Successful commercial deployment of storage thus had immense technical significance. In engineering, actually financing, building, and selling a thing have vastly more gravity than simply imagining it. It has gravitas. Moreover, the beauty of thermal storage is that it has virtually unlimited expansion capacity. One simply gets bigger tanks and puts more salt in them. Even though Andasol-1 had only seven hours of storage, it provided a credible precedent for storing the power of great cities for months, as necessary, and thus for providing the surge capacity required for powering Europe.

It is thus shocking how little this event registered in America, even as word of it raged through Europe like a fire. It was as though the U.S. technical press had set out to prove all those awful things Europeans are fond of saying about Americans' inability to understand anyone but themselves. Early in 2007, almost a year after Andasol-1 had broken ground, false reports began surfacing that thermal storage would be used in the Nevada Solar One project near Hoover Dam.[369] Six months after that, *Technology Review* ran a story about a solar startup (now pushed out of the solar farm business) and mentioned rumors that one plant in Nevada and two in Spain near Granada (presumably Andasol-1 and Andasol-2) used molten salt.[370] Two months after that, the *Wall Street Journal* reported the creation of a new company to make power plants using molten salt technology, again without any mention of Andasol.[371] Three months after that, the *New York Times* ran a story about how this company's technology would work, again without mentioning Andasol.[372] One month after that, the *Los Angeles Times* ran a story along the same lines, explaining that the technology in question had originated at Solar One, the remains of which could be seen near Barstow, again without mentioning Andasol.[373] Six months after that, the very week that Andasol-1 began delivering electricity to the grid (and one month after the Andasol story had broken in the European press), *CNET News* reported that an American company had partnered with Spanish company Preneal to build a fifty-megawatt solar energy plant one hundred miles south of Madrid, again with no mention of Andasol.[374] A week after that, a crew began demolishing the remains of Solar One.[375] Four weeks after that, the *New York Times* ran a story about salt storage, again with no mention of Andasol.[376] The story finally broke in the United States two months later by way of an article in *Scientific American*.

Events like this bring to mind the famous Henny Youngman joke: "I don't know what's so great about those German scientists. They're no better than our German scientists."

The salt of choice in such plants is not conventional table salt, which has an inconveniently high melting temperature, but a eutectic mixture of sodium nitrate and potassium nitrate known in the trade as solar salt.[377] The familiar example of a eutectic is solder, a mixture of tin and lead that melts at a lower temperature than either pure tin or pure lead do.[378] (By interesting coincidence, solar salt's melting point is about the same as solder's.) Nitrate salts are notorious oxidizing agents—and a key component of gunpowder for this reason—but they are nonetheless surprisingly stable in an environment containing only hot steel pipes.[379] Their main decomposition mechanism, oxygen loss, turns out to be beneficial, in that the nitrite ions left behind lower the salt mixture's melting temperature still further. Nitrates also have twice the heat capacity per unit volume as table salt, on account of packing more atoms into a given volume. They are also environmentally friendly. The nitrogen in the fertilizer we buy for our roses at the nursery is mostly nitrate. Sodium and potassium are common in the sea and also essential to life. Obviously it isn't a good idea to put a million tons of solar salt into the local groundwater, but a small amount spilled on the ground won't harm a thing.

Hot matter's ability to store energy is immense. Just as a kitchen toaster would take a long time to heat up a small Olympic swimming pool, so too would the toaster glow a long time if powered by the heat of the pool. The amount of energy we store in the pool by raising its temperature from the freezing point of water to the boiling point of water is enough to power the toaster for thirty years.[380] If we now enlarge the pool somewhat by doubling its length and increasing the width and depth to equal this length, thus making a huge cube of water one hundred meters on each side, the energy will power the toaster for thirteen thousand years—or all five counties in the Los Angeles Metropolitan Area for eight hours.[381] Molten solar salt's heat capacity is about two-thirds water's, and the temperature range exploited in practical applications of solar salt is about the same as water's, so the cube of this size would power Greater Los Angeles for about five hours.[382] More to the point, a farm one kilometer (0.62 mile) on each side packed with large petroleum refinery tanks filled with hot solar salt would power Greater Los Angeles for three days.[383]

Storing energy as heat is not that simple, unfortunately, because heat energy is accompanied by entropy, that bane of power engineers that, like junk mail, cable television channels, and official hot air, never decreases. Creating entropy is like gaining weight. One can always get rid of the extra, but doing so involves pain. In the case of thermal storage, the pain takes the form of energy we must throw away. The only way to get rid of entropy is to slough off heat at a low temperature. Thus, if we heat up the salt with electric heaters and then retrieve the energy back out with a steam turbine, we find that we can only get about one-third back. The rest has to be dumped out into the environment through cooling towers. Solar thermal power plants don't have this difficulty because they collect the sun's energy as heat in the first place. It makes perfect sense for this kind of plant to park the heat temporarily in a salt bath and then extract it later as needed. But other kinds of solar plants (and also wind plants) make electricity directly, so storing their energy as heat is much less straightforward. It cannot be done profitably at the moment.

However, it is theoretically possible to store electricity as heat without any loss. It requires only something called a heat pump—basically an engine so well constructed that it will run backward and function as a refrigerator. We don't routinely encounter such engines because market considerations usually require manufacturers to optimize for one task or the other. Thus, we find lots of machines that extract energy by moving heat from hot to cold (an automobile motor) and lots of machines that use energy to force heat from cold to hot (a refrigeration unit), but almost never do we find machines that will do both. However, perceptions can be misleading, for these two functions are actually mirror images of each other. If they're artfully constructed, neither creates entropy, so they're also functionally the reverses of each other. Accordingly it is possible, say, with turbo machinery, to pump heat from a low-temperature thermal reservoir (for example, a water tank) up into a hot salt bath, thus depositing more energy into the salt than was consumed in electricity. The same machine working in reverse then extracts the energy back out, necessarily dumping much of it as heat back down into the lower reservoir. This is fine, however, because the dumped part exactly equals the amount pumped uphill in the first step. No entropy is created, and all the energy put into the bank comes back out. However, no heat pump of the size and efficiency needed to achieve such storage presently exists.

At the moment there are two main approaches to extracting energy from the desert sun. One is to concentrate the sun's rays with mirrors onto heat collectors, which then generate steam from driving electric generators with turbines. The most successful of these so far is the parabolic trough, which concentrates sunlight into pipes filled with heat transfer fluid, typically a eutectic mixture of biphenyl and diphenyl oxide, that then flows to the steam boilers. The other is to extract energy from sunlight, either concentrated or not, with semiconductor solar cells. The semiconductor approach turns a greater fraction of the incident sunlight into electricity, but it is more expensive per given amount of power—at least in the desert, where land is plentiful and cheap. Trough technology thus holds a commanding lead. Neither technology contributes significantly to the world's present energy supply, however, and neither is cost-competitive with the wind.

Nonetheless, an energy boom based on these two technologies is presently underway in the Mojave.[384] Scattered about in the desert around Las Vegas but also stretching northward toward Tonopah, westward toward China Lake, and southward toward the Chuckwalla Valley are new solar power plants, both planned and under construction, totaling somewhere between four and fifteen billion watts.[385] It's difficult to give more precise numbers because solar power companies come and go relatively quickly, and the information about them is not always correct, for this is Wild West capitalism at its most earnest. If California Energy Commission estimates are correct, the power figures imply capital in play of between $15 billion and $500 billion. The company leading the pack, interestingly enough, is the one that built Andasol.

That the desert should be booming with a technology so expensive is highly significant, for it implies that investors are betting on future trouble in the natural gas markets and see sufficient advantage over wind to justify the extra cost.[386] Government subsidies are a factor too, of course. One of the advantages direct solar has over wind is reliability. The sun beats down on the deserts like clockwork all summer, and it does so most fiercely in the afternoons when people in nearby Los Angeles want their air conditioners on. But the more important advantage is that direct solar energy is more plentiful. The nearby Tehachapi wind area, the second largest in the world, has the potential to generate about five billion watts.[387] The entire Mojave, by contrast, has a potential to generate five *hundred* billion watts, even after discounting for 3 percent solar plant efficiency.[388]

Thus it isn't just our overactive imaginations telling us that Los Angeles might become the world's first great solar city. It's actually happening. When faced with developments of this nature, friends in Europe often joke, in an envious sort of way, that the aliens from outer space always seem to land in America first. The aliens don't always land in America first, of course, but it's difficult not to smile at the suggestion that they do and think that there could be no more appropriate place than the City of Angels.

Vegas is nearer many of these solar fields and could, in principle, become solar before Los Angeles does. It probably won't, however, because it's a night city that exists only when the sun isn't shining. This is, of course, why we go there.

Solar energy's extraordinary growth can be expected to slow, even after accounting for the effects of recessions. The renewable energy industry is partly a creation of government, and such institutions are notoriously unstable. The voters might very well lose their enthusiasm for solar energy once its true costs began showing up on their electricity bills. Lots of people in the financial community think that solar energy won't survive unless it reduces its capital costs by a factor of three.[389] Extreme pessimists think even that might not be enough, because the coal and natural gas industries can be expected to defend their businesses if threatened. But solar technology is new, and cost reductions of this magnitude sometimes occur in young industries as they discover new economies of scale.[390] If solar energy's costs did come down by these large amounts, the bets on the table could pay off spectacularly, for the new industry wouldn't stop with just powering Los Angeles.[391] But it's more likely that solar energy's costs won't come down sufficiently, in which case the payoff will be encouraging, but not spectacular, and the industry will continue to require government protection until coal and natural gas supplies begin to dry up for good.

Solar energy might need government protection longer than that, unfortunately. If nuclear energy hasn't disappeared by then, it's likely to be fiercely competitive and fond of squeezing its opposition the way natural gas does now. Nuclear energy is so politically incorrect at the moment, it's almost impossible to think of it as a proxy for natural gas, but that's precisely what it will become. It's actually slightly cheaper than natural gas at the moment, just not by enough to matter.[392]

The coming conflict between nuclear energy and solar energy has the particularly ironic twist that the two are fundamentally the same thing. For all its majesty and poetry, the sun is really just a big nuclear reactor in the sky that has gotten so hot it glows. It's a very well built reactor, one that has worked reliably and productively, as far as anyone knows, for over four billion years, but this is mostly a stroke of good fortune.[393] Other, inferior, models blow up spectacularly now and then, creating much entertainment for us and, presumably, very serious health problems for anyone living near them.[394] Like reactors here on earth, the sun burns fossil fuel, albeit of cosmic origins. Like reactors here on earth, it creates nuclear waste and sequesters this waste on site. Like reactors here on earth, it produces heat and nothing else. The blinding light streaming down to us may seem exotic, but it's actually just an especially hot version of the warmth of a steam radiator or the red heat of a fire's embers. If the sun were to go out, we could re-create its light, in both quality and quantity, simply by paving the moon with white hot bricks.[395] The bricks would cool off in about a minute and stop glowing, of course, if they weren't continuously heated from below, so mighty fires of some kind underneath would be needed to keep the energy coming—just as they are in reactors here on earth.[396]

The sun differs from terrestrial nuclear reactors, however, in transmitting all its heat through radiation. This is actually good, in that radiation travels at the speed of light and requires no medium, such as air, to propagate. That makes it ideal for traversing vast stretches of space. Radiation is also the optimal pipeline for transporting energy at white heat temperatures. An incandescent lightbulb filament, for example, gives up nearly all of its heat to radiation and practically none to its supports and electrical contacts, because the radiation channel is, in effect, a bigger pipe.[397]

But the price we pay for transporting the sun's heat radiatively is that the energy arrives spread out over large areas and thus requires a large, expensive apparatus to harvest. A very large nuclear reactor might funnel all its energy as hot water through pipes about as big as your front door in total cross-sectional area.[398] The sun, by contrast, spreads this same amount of energy over one million front door areas.[399] The size of solar plant needed to capture this energy winds up being even bigger, about one hundred million front door areas, after we reckon in all the engineering practicalities.

Solar energy's voracious appetite for land is so great that it's very likely to become a major environmental issue as the fossil fuel depletes. This has already happened, to some extent, with the political skirmishing now taking place over whether this or that public desert land should be given over to the new industry or protected as parkland or wilderness.[400] However, the skirmishing is just a foretaste of what's likely to happen after the plants are actually built and the full implications of their future proliferation become clear. This assumes that they succeed economically, of course, since the problem will disappear if they don't. The disagreements are likely to be about endangered tortoises, very special weeds, and the like, but they're really much more than that. To many people, particularly those who choose to live in the desert, the great value of this country lies in its vastness and emptiness, something we subvert by siting fleets of enormous solar power plants there. To paraphrase John Muir, filling the desert with power plants would be like filling a cathedral with them. To others, particularly those zooming by on their way to Vegas, who don't care about wilderness in general and the desert in particular, the entire proposition of protecting such land and its miserable little animals is preposterous. A meeting of minds on things of this nature isn't something we can expect to happen any time soon. The conflict could conceivably be more intractable than the present-day nuclear one, since solar power plants, unlike nuclear power plants, really do degrade the environment.

Moreover, the international version of solar energy's land problem is likely to be even thornier than the domestic version. It's one thing for your countrymen to despoil your land for profit, but quite another for foreigners to do this. Were the commercial development to become extensive and sufficiently lopsided in favor of those foreigners, people might be moved to expropriate the foreigners' property, ostensibly on the grounds that the foreigners were unjust exploiters who needed to be taught a lesson, but really just because they could. This has happened repeatedly in the oil industry, and there's no reason to expect it wouldn't happen in the solar energy industry just as readily.

Accordingly, solar energy is unlikely to supplant nuclear energy as the most important source of electric power in the world after the coal runs out. The immediate reason is that solar energy will have great difficulty falling in price sufficiently to match nuclear energy's cheapness, particularly since nuclear energy will itself be getting cheaper. But a subtler, and probably more important, reason is that solar power will still have land use and national security issues that are extremely serious, and it will still not work in parts of the world that are dark. Europe will never be completely powered by the Sahara.

Nonetheless, for those parts of the world that aren't dark and are endowed with lots of sparsely populated desert, solar energy is likely to become a major source of electric power and possibly even the sole source. The American Southwest, for example, will very likely be powered by the sun after the coal is gone, as will Western Australia, northern Africa, and the Arabian peninsula. Solar electricity is also likely to become an export commodity from those regions, although the amount of export will depend on future long-term storage capacity in the importing regions and countries.

The truly important consequence of solar energy's rise in the next two centuries is thus likely to be economic transformation of the world's deserts. Las Vegas is a particularly notorious example of the rapid growth that can occur when these hot, dry places are supplied with cheap power, but there are plenty of others, particularly in the Persian Gulf. Deserts are actually quite desirable places to live once you have electricity. The pollen-free air is good for health. The heat is pleasant, like a sauna. When you tire of it, you take a dip in your swimming pool (filled with desalinated water), after which you towel off and disappear into your air-conditioned house. Even if you imagine a desert covered with great facilities making power for export staffed by relatively few people, it's just a matter of time before supermarkets spring up to serve those people, followed by schools, highways, auto repair shops, restaurants, and all the other components of suburban life. Energy-intensive industries, such as aluminum and electric steel, will soon follow, lured not just by cheap power but also by cheap land and labor. Light manufacturing—deck furniture, household appliances, building materials—won't be far behind, for the same reason. And so on and so forth. The towns and cities that spring up will obviously have a different cultural flavor depending on whether they are filled with mosques, casinos, or Mormon temples, but the underlying economics will be the same.

However, even if nuclear power wins the competition and comes to dominate world energy production when the coal exhausts two centuries from now, solar power will be there in the background, ready to step into the breach and supplant nuclear power any time the world wishes. That might be a few decades from now when governments ban nuclear power, a few millennia from now when the nuclear fuel supplies tighten, or never, depending on circumstances. But exactly when the transition occurs is, in some sense, a detail. The truly important thing is that solar energy will be a reliable insurance policy, a guarantee that the world never starves for energy so long as the sun shines, which, as far as anyone knows, will be a very, very long time.

CHAPTER 10

Under
the Sea

The bottom of the ocean is a hellish place. The water there couldn't get any colder without turning into ice. It's melt water from the polar ice sheets that sank to the abyss many centuries ago and got stuck there.[401] Sunlight cannot penetrate to such depths, so the ocean bottom is blacker than night, except for the glow of an occasional light-producing fish passing by.[402] The pressure there is four hundred atmospheres, enough to crush air to half the density of liquid water and to prevent hot water from boiling at any temperature.[403] Human beings will always be fascinated with the ocean depths, but they will never live and work there physically because it's too hostile an environment.

Nonetheless, the deep ocean is likely to become increasingly important for the world economy—and the energy sector in particular—over the next two centuries as fossil fuels deplete.[404] The near-term reason is that the world's large oil resources under deep water at the outer continental shelf require much trouble and expense to extract, so they will matter more and more as wells elsewhere fail.[405] The daring deeds of oil drillers in the Gulf of Mexico and the North Sea are currently being repeated off the Atlantic coasts of South America and Africa, and even more astonishing feats in the Arctic Ocean will soon surpass them.[406] There are also extensive undersea methane hydrate deposits that energy companies may very well mine when natural gas eventually becomes scarce.[407] But even after all these oil and gas resources are gone, the ocean bottom will still be important as a source of geothermal heat and a place where we can store things, one that won't conflict with human land use needs because humans will never live there. It has the potential, like the world's great deserts, to become increasingly valuable real estate even though it has no such value at the moment.

The facilitator of this economic rise of the ocean deeps will be the robot. More precisely, it will be the Remotely Operated Vehicle, a machine with all the mechanical parts of a robot but piloted remotely by a human operator, like a puppet.[408] The communication of video and tactile information to the pilot as well as commands back from the pilot are by optic fiber and wire telemetry, so the pilot can sit in a ship above the work site or on nearby land, where it's safe, and work from there. In effect, the body is in one place while the brain is in another. It's not so different from the flight simulators we run on our computers, except that what's on the other end is real rather than fictitious.[409] Genuinely autonomous machines that think and act on their own will be present too, but they probably won't be doing sophisticated things because they probably won't have gotten much smarter than they are now. Exactly why artificial brains are so difficult to make is a matter of debate, but the sixty years of failed attempts make this debate academic. The oceans are presently swarming with autonomous robots shaped like cruise missiles that swim around on preprogrammed trajectories and make surveys.[410] They do their jobs well, but they are nonetheless the intellectual equivalents of microwave ovens. Meanwhile, hundreds of robot puppets work at the bottom of the ocean carrying parts about and bolting them together at the bases of deep drilling rigs and well heads in the service of oil companies. This activity is indispensable to drilling for oil in deep water, so the numbers of such robots will very likely increase over time.

Robots can live and work in this terrible place for the simple reason that it isn't terrible for them. The pressure even in the deepest parts of the ocean is negligible on quantum mechanical scales, so it has no effect on the metals, plastics, and fluid oils that make up a robot's body. Swimming around on the ocean bottom is no more difficult for a robot than swimming around in a backyard pool. The icy temperature doesn't matter either. If anything, its constancy is helpful because weather of any kind is an opportunity for trouble. Very heavy robots especially like the deep water because it adds buoyancy that helps them lumber along the bottom on their tractor treads or skis.[411]

Also, the robots aren't lonely down there at all. The wire and optical fiber communications with their puppeteers two kilometers above travel at the speed of light, like telephone signals, and are comprehensive. The robot sends sophisticated sight, sound, and motion sensations to the operator. The operator returns sophisticated instructions about where to look, how to move, what to grab, whether to twist gently or forcefully, and so forth. Thus the communication system is rather like a spinal cord. Computers even mediate, coordinate, and modify signals in both directions the way the lower brain does. The robots are, for all practical purposes, people who have come down the wires temporarily to work on oil jobs in pressure-tolerant mechanical bodies.

Becoming a robot is expensive at the moment, of course. A low-end machine that lacks grappling arms and can swim only as deep as a scuba diver (one-tenth the depth needed for oil work) costs about as much as a mid-size car. A serious machine that can go to great depth and has arms, sensors, and customized tools costs about as much as a median-priced home in Beverly Hills. We also need a topside vessel and crew, which for deep diving means a multi-million-dollar ship. The present-day robot population is thus small and limited to uses for which their cost is marginal, such as drilling for oil, repairing undersea power and communication cables, clearing mines, and exploring sunken wrecks.

However, the cost of mechanical bodies is likely to fall dramatically in the future. One reason is that high-technology gadgets tend to do just that, as anyone who has bought a television or laptop computer recently can attest. An equally important factor, however, is that robot price warfare is likely to break out as oil supplies finally fail for good. Exactly the opposite will happen when oil first begins to tighten, of course. Energy prices will skyrocket, and that will increase demand for undersea drilling robots and, thus, also their prices. But that demand will also stimulate robot companies to expand their production capacity and improve their technologies. When this extra demand eventually flags as oil depletes further, these companies will be so over-built and overdeveloped they'll have no choice but to cut prices to the bone and flood the world with cheap robots.

Exactly what will happen when robot prices collapse is difficult to predict, for the scramble to exploit this newly cheap resource will very likely unleash an orgy of innovation, just as the collapse of semiconductor memory prices did in the early 1980s.[412] The result in the latter case was a parade of things so fabulous that our grandparents could not even have imagined them: the word processor, the graphical operating system, the Internet, the web browser, the laptop, the scanner, the digital game, the digital camera, the digital newspaper, the digital book, the intelligent cell phone, and so forth. It's also possible that robot technology will simply disappear the way steam locomotives, vacuum tubes, and manual typewriters did when nobody needed them any more, but it isn't likely.

Thus, one thing we can predict with reasonable confidence is that humans will use these robots to colonize the deep ocean. It won't be the Jules Verne kind of colonization with submarines, deep diving suits, air locks, and great underwater cities.[413] That will still be physically impossible. Rather, it will be with robot doppelgängers that people control from the safety of their workplaces and homes on dry land. The number of these machines populating the deep will depend on the economic opportunities for them waiting there, but it could conceivably reach billions, for it is has no fundamental limit other than the number of individuals back on land willing to be pilots. That could be extremely large, judging from the number of individuals back on land presently willing to play computer games. A great population of undersea robots would be conceptually the same thing as a present-day massively multiplayer online role-playing game, except that there would be real physical agents on the other end instead of avatars and real money at stake instead of play money.[414]

Invasion of the oceans is especially likely, in part, because no sophisticated invention or scientific discovery is required to make it possible. The only missing ingredient is a subsea infrastructure of roads, power lines, and communication lines, a matter of money and drive for sure but not a matter of cleverness. Undersea robots will always require power and communications cabling, for they need energy to operate, just as fish do, and high-capacity data links back to their operators. The latter is an especially firm constraint because water's physical properties prevent us from transmitting information through it at high rates over distances greater than a few tens of meters. One or two robots working on an undersea oil rig are reasonably provisioned with cables from a mother ship above. Large teams of robots, by contrast, aren't cost effective unless they're linked by cables laid on the ocean floor to a central facility of some sort. Teams of teams then require cabling of these facilities to even more centralized facilities, and so forth. The end result is a great tree, the trunk of which emerges onto land somewhere, thus eliminating the need for a billion ships. The robots themselves—or more precisely the people piloting them—have to construct and maintain this tree, for the environment is too hostile for humans.

How all these robots might earn a living requires a bit of imagination to see, because the subsea economy doesn't yet exist. However, some speculation is appropriate here because much of the post-peak-oil land economy probably doesn't exist yet either. Undersea telecommunications companies will very likely have grown substantially by that time, if only because higher fuel prices will have made travel more expensive. Expansion of their businesses will call for a lot more robots for constructing and maintaining their cable pathways. Tight energy markets will probably induce electric utilities to lay undersea power cables over great distances, perhaps even between continents. These will require not only ready robot repair crews but also permanent robot staffing of undersea power consolidation and distribution centers. There is likely to be an increase in undersea trash disposal, notwithstanding present-day abhorrence of the practice, an activity requiring not only transport of waste to the disposal site but also burial with robotic soil-moving equipment. Teams of robot workers capable of aligning pieces of steel and bolting them together would be needed to construct intercontinental pipelines suitable for transporting freshwater or industrial commodities over great distances without pumps. Robot teams armed with vacuums, shovels, and perhaps also rock drills would be similarly needed to mine the ocean depths for the minerals known to be there.[415] Robotic transporters would be needed to ferry crews over large distances from one work site to another or to centralized pickup points for repair. Robot hospitals, repair shops, and parts distributors would be needed to circumvent expensive transport of sick robots back to land. And all of these activities would be vulnerable to disruption and sabotage, just as the corresponding human activities are on land, so there would have to be a robot police force to impose order, robot pounds in which to incarcerate offenders and watch over lost or stolen property, and robot armies to defend national interests.

Robots will also be gathering energy under the sea, even after discounting the huge oil and methane deposits there. The ocean floor along tectonic plate–spreading boundaries is an especially potent source of geothermal heat, a fact that manifests itself through spectacular chains of underwater hot springs along mid-ocean ridges.[416] The grandest of these, the famous black smokers, pour out water at temperatures and flow rates characteristic of small nuclear reactors.[417] The total geothermal heat released in such springs in all the world's mid-ocean ridges is slightly less than civilization's present-day energy consumption.[418] Robots skilled at drilling would be well suited for tapping these hydrothermal vents for use in powering undersea power plants.[419] Drilling into mid-ocean ridges would have the additional advantage of minimizing citizen complaints about earthquakes let loose by accident (which commonly arise on land) because no people would be living near the drill sites.[420]

Robots will not, however, be defying the laws of physics. They won't, for example, be extracting the world's energy needs from deep-ocean currents using turbines.[421] There isn't enough energy in the currents to extract. Even if the entire ocean were moving at the speed of the Gulf Stream, which it isn't even close to doing, the amount of kinetic energy it contained would supply the world's present energy needs only for about six years.[422] Taking this energy out would also be a very bad idea, for it would stop the world's thermohaline circulation and change the world's climate. The robots also won't be piping warm surface water down to the ocean depths and cold water back up in order to generate electricity from the heat difference.[423] The Carnot efficiency of this scheme is so terrible that the flow of water required to power the world, even with thermodynamically perfect engines, would be four hundred thousand times the maximum flow of the Alaska Pipeline, or eight Amazons.[424] The robots, or their topside accomplices, won't be supplying the world's energy from the tides either.[425] The amount of energy the earth-moon system continually deposits into the tides, known from measurements of the small but inexorable lengthening of the day, is less than one-third of the world's energy needs.[426] Also, if we extracted all this energy, the tides would stop.

The robots will also very likely be storing some of the world's energy in the abyssal depths. Exactly how much depends on relative costs, which are difficult to assess before undersea infrastructure is built. However, storing energy under the sea has some potentially important advantages over storing it on land. The facilities would be immune from weather and easy (in principle) to build, inspect, and repair. Like underground facilities, undersea ones wouldn't use up valuable land, create eyesores, or endanger people.

By far the simplest way to store energy under the sea is by bubbling compressed air into tanks or bags. We might, for example, tow a big steel oil refinery tank to the appropriate spot, fill it with water so it sinks to the bottom, and then punch many openings at the base so that seawater could flow in and out. Air pumped down from the surface and piped into one of these openings wouldn't actually bubble up inside, because air at abyssal pressures has about half the density of water. It would instead form globs, the way gasoline does, and float up to the top of the tank in a leisurely sort of way, displacing water downward as it did. An enormous amount of energy would be required to compress surface air to this density for storage, just as is the case in scuba tanks, the pressure of which is comparable.[427] We'd get it all back, however, at least in principle, simply by reversing the process and driving turbines with the high-pressure air. The energy stored in a given volume (after compression) of air at these pressures is roughly that stored in the same volume of molten solar salt.[428] The tank requirements would thus be comparable to those of thermal storage facilities. The pressure and weight loads on its walls would nonetheless be modest—about those we'd get if we filled the tank with liquid natural gas on land.[429] They'd just be upside down because they'd be coming from upward buoyancy instead of downward gravity. It would thus probably be more sensible for the robots to build tanks in place on the ocean floor using concrete and steel, along the lines of present-day natural gas terminal tanks, instead of towing in prefabricated ones. Big natural gas tanks are not only wide but also extremely tall—about half the height of Hoover Dam.[430] Underwater tanks could conceivably be even larger because their materials would effectively weigh less, and there would be no need for heavy thermal insulation.

Storing serious amounts of energy as compressed air on the ocean bottom would require many such tanks but not a lot of real estate. The Los Angeles Metropolitan Area, for example, could be powered for a day by a field of about six tanks with present-day dimensions.[431] All the energy in Lake Mead (which would power the Los Angeles Metropolitan Area for two months) could be stored in a field of about four hundred such tanks. That would require a square patch of ocean bottom about two kilometers on each side—an area two hundred times smaller than Lake Mead's.[432] A year's supply of electricity for the entire world could be stored in a field of about three hundred thousand tanks, or a patch fifty kilometers on each side.[433] A year's supply of energy of all kinds for the world could be stored in about two million tanks, on a patch one hundred fifty kilometers on each side. That is the size of New Hampshire or Israel[434] and less than one ten-thousandth of the available ocean floor.[435]

The stupendous amount of land available under the sea is the chief reason storing energy there might wind up competing economically with storing it underground in caverns, the preferred method at the moment.[436] Suitable caverns are relatively plentiful, for the oil industry creates them when it pumps wells dry, and they have the obvious cost advantage over underwater tanks that we don't have to build any tanks. We do, however, have to drill into the caverns, and that cost is difficult to assess because it's presently piggybacked on oil exploration budgets. That's important because, even at enormous abyssal pressures, compressed air contains sixty times less energy per unit volume than compressed natural gas does, on account of not being flammable.[437] A cavern volume adequate for natural gas storage is thus badly inadequate for storing energy as compressed air. The U.S. Strategic Petroleum Reserve, for example, which resides in deep underground salt caverns, has a volume equivalent of 150 large undersea compressed air tanks. That's less than half the number required to store the energy of Lake Mead.[438] The entire underground natural gas storage capacity of the United States has a volume equivalent of 750 tanks, just twice the number required to store the energy of Lake Mead.[439]

Ocean storage of energy would also be safe, notwithstanding the three atomic bombs of energy contained in each tank.[440] The reason is that the energy resides not in the air but rather in the air's tendency to cool and extract heat from the environment as it expands. Thus, if the air gets out of control and expands too fast, it doesn't explode but simply refrigerates its surroundings. Were a tank's roof to fail catastrophically, all the air inside would simply blub slowly upward, cool, and begin freezing seawater into ice. By the time the snowy mess reached the surface, nine-tenths of the energy would have gone to making a medium-sized iceberg, except chopped into bits in a slurry rather than one big chunk.[441] The rest would result in a violent roiling of the surface waters. The amount of agitation would probably be similar to that of the notorious 1987 Lake Nyos overturning event in Cameroon, which produced a similar amount of gas.[442] The Nyos event wasn't witnessed, unfortunately, because it occurred at night, but it sloshed waves twenty meters high on a nearby shore and probably produced a fountain of water and froth ninety meters high. An undersea tank failure would thus be an unhappy experience for ships immediately above and presumably for sea life in the way of the rampaging air, but it would have no deleterious effects beyond a few kilometers from the epicenter. But such a mighty breach would be unlikely to occur very often because of the tank's small overpressure. We'd mainly expect small leaks, which would result in an innocuous burping of the ocean surface from time to time accompanied by a few ice cubes bobbing up.

The turbines and compressors required for storing energy under the sea and extracting it again would not be safe, however, unless they also resided in deep water. In principle we could pipe air from undersea tanks up to the surface and run generators with it there, but in practice the nuclear-scale explosive danger of a big pipe pressurized to four hundred atmospheres and capable of powering Los Angeles would be intolerable. Fortunately, although deploying complex generation machinery underwater would be costly, it wouldn't be any more difficult technically than building a fleet of submarines. In fact, it would be easier because there would be no need to create a low-pressure environment for the sailors inside. The chief problem would thus not be the machinery per se, but rather the logistics of getting the necessary heat in and out. The relative scarcity of heat at the bottom of the sea is, of course, what makes storing great amounts of energy there safe.

The need to manage heat flow would probably constrain undersea energy tank farms, if they were built, to be located just beyond the continental margin. The reason is that the continental slope is a plateau escarpment, like the Drakensberg of South Africa or the Cedar Breaks of Utah, on which we can build nearby support equipment, such as heat exchangers and auxiliary storage tanks, at any depth we like simply by siting them properly.[443] Locating the tanks at the margin would of course also minimize the length of pipe and cable required to reach them from shore.

Why the slope matters is that we'd need it to extract the energy gradually in steps to avoid icing problems with the machinery. Suppose, for example, that we ran compressed air at abyssal pressure through a turbine and then piped it to a second tank two hundred meters higher up the slope. The pressure drop between the two tanks would be about the same as that between the top and bottom of Hoover Dam, which is just this height, and the working fluid, compressed air, would be about half the density of water, so these would be a reasonable set of operating conditions for electricity generation. However, passing the air through the turbine would cool it with this arrangement to about the freezing point of water, so the second tank couldn't be placed much higher without risking ice formation in the machinery.[444] In a barebones design, we would pass the air through a heat exchanger (a building-sized automobile radiator), thus bringing it back up to ambient temperature of the water before piping it to the second tank. We could then chain together about fifty such stages, each higher up the slope than the one before, to extract all the energy available in the original tank. Equivalently, we could just have fifty copies of the first stage at depth.

The amount of machinery we'd have to locate at depth is very large. For example, even when broken down into fifty stages, powering the Los Angeles Metropolitan Area would require about ninety-three tons of air per second (one-quarter of the water flow through Hoover Dam) to flow through each stage's turbines.[445] It would also require about sixteen tons of warming water per second (the water flow through the Los Angeles Aqueduct) to flow each stage's heat exchangers. That, in turn, would require the heat exchangers to be about thirty meters across to accommodate the relatively slow natural currents at the ocean bottom.[446]

The slope is also crucial for efficient energy deposition, although for a different reason. Heat only flows spontaneously from hot things to colder ones. Some of the heat of compression generated in the storage process is therefore lost forever and cannot be retrieved. One minimizes this loss by minimizing the temperature difference between the hot body and its cooler surroundings in the heat transfer process. Such minimization is achieved automatically when we exactly reverse staged extraction. By compressing the air only a small amount each time and sloughing off the heat thus generated before compressing to the next stage two hundred meters below, we can reduce the total round-trip energy loss, at least theoretically, to about 3 percent.[447]

There are other ways of regulating heat flow to safely manage enormous amounts of energy stored as compressed air in the deep ocean. One could, for example, pipe warm water down from the surface and expand compressed air by using its heat rather than with the small amounts available in icy bottom waters. That would reduce the number of required expansion stages to ten, but it would also consume lots of pumping energy and require a monstrous intake off Malibu or Catalina Island steadily slurping six Los Angeles Aqueduct flows (one-third of the Colorado River flow) of surface water.[448] Surfers, seals, and playful otters venturing too near such an intake would never be heard from again. It's also possible to save the heat of compression in big underwater tanks for later use in expansion. Abyssal pressures are so great that even extremely hot water, like that issuing from black smokers, could be stored this way without explosion danger. The heat required to expand a single undersea tank could, for example, be stored in an identical tank filled with boiling water—more precisely, with water raised to sea-level boiling temperature, for water at these depths won't boil.[449]

There are also ways of storing enormous amounts of energy in the ocean that don't involve heat at all. For example, we could just pump saturated brine uphill. Very salty water is denser than seawater, as anyone knows who has swum in the Dead Sea or Great Salt Lake. It will thus flow downhill of its own accord, even underwater, given the chance. Thus, if we dump saturated brine on the sea floor quickly, so that it has no time to mix with the surrounding water, it will puddle in local depressions and form underwater brine lakes. This wouldn't ever happen in nature, of course, because we couldn't dump the brine fast enough. What does happen in nature, though, is that depressions with large salt outcrops on their floors slowly saturate the water they contain and create brine lakes all by themselves.[450] Several such lakes have been discovered in deep waters of the Gulf of Mexico, the Eastern Mediterranean, and the Red Sea. A particularly large one in the Gulf of Mexico's Orca Basin, about two kilometers down, is as deep as Lake Mead and has a surface area about one-fifth Lake Mead's.[451] We could, in principle, build a pair of artificial lakes of this nature, one on the continental shelf and one on the abyssal plain, thus creating an undersea hydropump storage facility. To store energy, we'd pump brine from the lower lake against gravity to the upper one. To extract the energy, we'd drop the brine back down to the lower lake through penstocks and underwater generators.

The energy storage potential of undersea civil engineering projects such as brine lakes is large. The amount of energy stored for a given elevation difference and pumping volume would be about one-fifth that stored on land, because what counts is the density difference between saturated brine and surrounding seawater, not the density of the brine itself. But the enormity of the continental shelf elevation drop would compensate for that deficiency. A working pair of underwater brine lakes the size of Lake Mead, one on the shelf and one on the abyssal floor, would store four times the energy of Lake Mead.[452] We might think about increasing this to twenty times by filling the lakes with heavy cesium chloride brine, the trick oil drillers use to make dense drilling muds, but this would be a bad idea because cesium is an environmental menace and also would cost eight thousand trillion dollars per lake pair.[453] About two hundred conventional sodium chloride brine lakes of this size could store all the world's electricity for a year.[454]

Exactly when mass exploitation of the deep ocean will begin in earnest is difficult to say precisely, for it depends on exhaustion of oil supplies, a notoriously slippery concept.[455] One thing we know for sure is that all the oil fields in the continental United States, including the North Slope of Alaska, are now in decline, as is the North Sea, and that the U.S. Energy Information Agency assesses these declines, and others like them, as offsetting all new discoveries of oil outside the Organization of Petroleum Exporting Countries.[456] We also know that new strikes in old fields like the Gulf of Mexico as well as genuinely new fields like Carioca in Brazil are either small compared to Middle East proven reserves or already accounted for as part of those reserves.[457] We also know that some potentially stupendous reserves, notably the Venezuelan oil sands, are missing from petroleum industry estimates.[458] Thus the answer has a large spread. The British Petroleum figures for present world oil consumption give a time to exhaust all proven reserves of about forty years.[459] If we add in the omitted Venezuelan reserves, this increases to seventy-five years.[460] The figures for Middle East oil production give a time to exhaust all Middle East proven reserves of about eighty years.[461] Thus the undersea era will probably dawn sometime around the turn of the next century.

CHAPTER *11*

A Winter's Eve

I t's a bitterly cold winter night, one of those bone-chillers when all is still and the moon sails above scraggly trees coated with ice, painting them blue. Warm light glows out of an occasional window, indicating that real people are inside, presumably cooking dinner, attending to homework, or burning time with their ever-more-clever electronic toys, but anyway snugly protected from the cold. Smoke occasionally curls from a chimney, telltale evidence of a toasty fire flaming in the living room grate, but it's fairly rare, for people are too tired after work to light fires. Nobody is out. The snow crunches under the tires of cars that pass by now and then, for there are icy patches the plows didn't reach until it was too late. Otherwise, all is quiet.

This winter has been truly awful, one of the fiercest in recent memory. Blizzards have howled in from Canada one after another like freight trains, exhausting county snow budgets very early in the season and generating even more hand wringing than usual over where the additional money will come from. Hardy cross-country skier types have been out enjoying the abundance, or at least saying they do, but everybody else has been stuck inside dreaming of spring. The world is warmer now than it was two centuries ago, so they say, but it's not clear that anyone informed the Canadians of this fact.[462] Either that or it's a nefarious plot to sell us more snow blowers.

It's rather depressing to realize how often you observe scenes like this on the way back from business trips. This is both because you're taking too many trips and because you're returning too tired from each one of them to do anything but observe. The Shinkansen to New York is very pleasant once you get on board, but the trip to the terminal is maddening, as are the tortures they inflict on you, such as queues, coughing children, bad restaurants, dirty seats, and ubiquitous televisions that won't turn off. By the time you've found the hotels, met the clients, projected the competence, and repeated the travel steps a second time in reverse order, you don't have energy left to do anything except marvel at the cold and wish you were home already.

The kids inside these homes have no idea what's in store for them, the little darlings. It's probably just as well, for they'd refuse to grow up if they did. The younger ones can't think about this now, thanks to getting shuttled back and forth among school, tutors, and sports events by indefatigable mothers driving regenerative minivans that sip gasoline but nonetheless cost a fortune to run. The older ones are off at college, enjoying their last bit of freedom before things get serious. Rumor has it that there was a hilarious retro nuclear protest party a week ago in which people came dressed either as manifestos or as nuclear reactors and had to chant slogans in order to get punch. The chaperones came dressed as nuclear waste. It was all in extraordinarily bad taste, but that's the way twenty-somethings are. Of course, as Mark Twain said, it's better to be a young June bug than an old bird of paradise.

Oldsters need amusement during the long winter months too, of course, but we have more wicked senses of humor than the kids do, so we get our entertainment from the financial news. A good place to start is always algae futures. Algae price fluctuations don't raise people's blood pressure around here the way corn does because this isn't algae country, and algae are inherently funnier than corn, which is already somewhat funny. You just can't think about the inner workings of an algae processing plant without smiling, for it's a malevolent green slime monster movie in all but name. If algae news doesn't cheer you up, pig and chicken manure gas spot prices definitely will, for it's even more difficult not to smile when people are talking with complete seriousness about benchmark Wilbur Sweet Crude, Colonel Sanders Light, and Amsterdam Blend. Your no-fail backup amusement of last resort is news from the Chicago Garbage Exchange pits, where disgusting guck of all imaginable grades and origins is bought and sold by people yelling as if the stuff were gold, which, to them, it is.

The light glowing from these windows is so brave and individualistic it's hard to believe sometimes that much of it is last summer's sun captured and stored up in various places. It's one of those modern miracles of science— like the ability of airplanes to lift off the ground and fly—that will never make sense intuitively no matter how many times they explain the facts to you and show you the equations. Exactly what fraction comes from the sun isn't clear, partly because it varies from one season to the next and partly because it's politically sensitive. A reasonable guess is about a third, but it could easily be twice that. However, any estimate you like, after adding up all the customers, amounts to several nuclear wars' worth of energy stored away and retrieved again every year, a concept that just boggles the mind.[463]

Nobody cares where their electricity comes from, of course. People only care that it should come when they wish and cost relatively little. It does both remarkably well, actually, although maybe this is not so surprising given the howl people would raise if it didn't. However, electricity has always been extremely cheap. Indeed, historians tell us that it's actually cheaper now than it was two centuries ago, an astonishing fact given how much more expensive fuel for cars is today. Electricity prices do rise slightly now during the Christmas holidays, the ostensible reason being to discourage irresponsible use of outdoor decoration lights. But the rise doesn't ever seem to do much actual discouraging. Even in a quiet neighborhood such as this one, the arrival of Christmas brings an orgy of bejeweled trees, dancing elves, and blazing Santa sleighs with Rudolph the Red-Nosed Reindeer. It's fairly obvious that the power people have megawatts to spare, and that they consider it a matter of honor to provide homeowners with however much they want, regardless of how irresponsible those wants might be.

A few of the people behind those windows are probably burning midnight oil on the Internet—although not many because firms rarely like their employees working from home on account of the security risk. Confidentiality breaches are a perpetual problem in electronic business, but they're particularly dangerous in remote robotic work, which a lot of people here do for a living, because they can cause not only communications mischief but also serious physical damage or even loss of life. That's presumably why land-based robotic operations are so much more highly regulated than undersea ones. A rampaging robot or two at your undersea power plant might blow up some tanks and terrorize the fish, but a few big ones above ground could destroy a city. Large-scale vandalism of this nature has never happened in recent memory, but there are reminders all the time of why you need to remain vigilant. Just last week somebody broke into a dairy computer and ordered all the milking machines to commit suicide. It was probably a prank, and no cows were hurt, fortunately, but if they had been it could have been quite ghastly.

Some people are also traveling electronically tonight. Physically going to exotic places, particularly ones overseas, is horribly expensive and financially out of reach for most people, but they can go there virtually by computer, and often do. The truly high-quality experiences obviously require the sophisticated equipment you only find in travel agencies, but the more primitive ones use home equipment and can be purchased for reasonable prices by credit card. The hot new adventure this year is apparently driving across the Pacific. You rent a robot in San Francisco, usually in collaboration with friends, for it's a grueling drive requiring shifts, and then pilot the thing remotely along the long road to Tokyo. Only individuals with lots of time on their hands can do this, for the pace is slow and the distance very great, between one and two times the distance across Siberia, depending on how you measure. However, they say you have interesting encounters down there, sometimes with strange fish but mostly with strange aspects of civilization, such as lonely undersea repair stops linked to casinos, frighteningly large freight transports, and buses full of robot work gangs buzzing each other with short-range light flashes and croaking. On occasion a person's robot gets waylaid and presumably sold for parts in Mexico or Russia, so you have to get insurance. When you're done you get bragging rights and a certificate saying that you've crossed the Pacific.

Folks trekking out in the ocean only rarely catch glimpses of the most interesting thing of all down there—the great long-haul power and communications cable bundles that stretch between the continents. They're extremely sensitive, for obvious reasons, and mostly off-limits to anyone not needing access to them. You see them mainly where a branch or backup line crosses the highway. The intersection is easy to spot because it's lit and patrolled. If you stop your robot there to gawk, officers will come and arrest it, and you'll have to pay serious money to bail it out of jail. The power of nations courses through those cables. About six of them can light up a big city like New York. Four hundred of them can light up Europe.[464] They crisscross the oceans along tracks between great undersea hubs where the flows of power they carry are brokered and channeled to various customers. The system never works as well as it should, thanks to the headaches of international politics, but it does work, even if only as a stabilizer of the world power system. Some enthusiastic fans even argue that it's the greatest engineering feat of history.

On very cold nights such as this you crave comfort and find it in the knowledge that the Canadians are sending down, in addition to frigidity, lots of uranium. Actually, they are also sending down lots of electricity made from uranium, so they're bestowing the benefit twice, but the uranium shipments provide more security, so they count more. Not everyone is happy about these shipments, of course, and there is political agitation against them every year. It's relatively muted, however, both because people don't like being cold and because the great time-scales involved are becoming steadily more tangible. The fission dirt of the world's original reactors is about one-quarter of the way through its aging process, and its slow march to benignness isn't just a theory anymore but a matter of public record. Whether authorities will release it back into the wild when the aging is done, as they presently plan to do, is an interesting question, for the amount of material involved is so small that it might be cheaper, after the political cost is reckoned in, to leave it sequestered and just enlarge the storage sites. Human nature is very consistent one generation to the next, so some people will always believe that the longer-lived radioisotopes weren't actually separated out, recycled, and burned up in reactors, as the experts claim, but were instead left in to save money. This is true even if the stuff generates no Geiger-counter clicks. Obviously if it makes clicks, the facility will get enlarged.

Although fission pollution isn't on most people's minds that much, other kinds of pollution definitely are, for the world's as crowded as ever, and the environmental consequences of this crowding continue to be dire. Every year ecologists say that the earth has lost more species, and it's never species you want to lose, such as poison ivy and cockroaches, but ones you don't, such as Amazonian plants with potentially valuable medicinal powers and exotic butterflies, rare weasels, and so forth. Big cities remain notoriously dangerous places to live, which is one of the reasons the houses here have lights in them. Urban poverty is still endemic, even after two centuries of efforts to eradicate it, and perhaps has even worsened slightly because the world economy left a lot of people behind when it made the transition from oil. The Greens and the manufacturers are constantly at war with each other over eliminating metals from the trash stream, particularly those in batteries. The environmentalists lobby every year for higher return deposits, and the manufacturers threaten every year to leave town if they're enacted. Last year there was another scary but ultimately unsuccessful campaign to shame people into not using their cell phones because they contained batteries. The year before that the target was mechanical toys. There's also constant maneuvering to get everything made out of carbon, the only substance many people say doesn't harm the environment when burned, even to the point of physical preposterousness, such as calling for plastic engines and ball bearings.

The environmental issue that isn't so preposterous and that's likely to escalate in seriousness as time passes is saving the world's deserts from being paved over with solar plants. Not everybody has a soft place in their heart for desert wilderness, of course, but many people do, and they are concerned that it's disappearing. Once a power plant begins generating profit, copycat plants begin springing up, whereupon somebody pipes in water, and voilà, you have a town. If the town is well planned, which it often is, people then flock there because it's a great place to live. They escape snow, pollen, and big-city headaches, and the baking heat is no problem because there's plenty of power for air conditioning. In the American Southwest at the moment there is an uneasy truce between desert developers and environmentalists in which most of the land is still wild. The truce is unstable, however, and could break out in ugly political fighting at any moment. There is no such equilibrium in the other great deserts of the world. All of them are growing explosively at the moment.

It's remarkable how primal night becomes as the temperature lowers. Your neighbors are just on the other side of those windows, but they might as well be on the other side of the moon. There aren't any wolves around here, but if there were they would definitely be howling. Why it's primal is that Nature is reminding you, ever so gently, who's boss. The amount of heat that flows in and out of Hudson Bay as it melts and freezes every year is greater than the entire energy budget of civilization.[465] The amount of heat that flows in and out of Canada generally is between ten and twenty times that, depending on how much of the polar ice cap you count.[466] Humans couldn't push back this cold even if they tried. They're not in its league. The same goes for the even bigger action in the tropics. The amount of heat that flows in and out of the world's oceans in making rain and snow is two thousand times the energy budget of civilization.[467]

You therefore have to hand it to people of the past, who made their way in this world where Nature is boss on a flaming log or two and, later, on the remains of plants and animals long dead and buried in the rocks. They must have been very resourceful. Today, we take the kids on field trips to the Coal and Oil Museum and encourage them to push the diorama buttons, but such things give you only a superficial understanding of the daring deeds performed in those days, the gambles taken, the fortunes made, the lives lost. A somewhat better understanding comes from your own prosperity, plus, of course, the atmospheric carbon dioxide records, which show that human beings transferred a trillion tons of carbon from the ground to the air during the last two centuries.[468] The petroleum component of this transfer amounted to the flow of the Mississippi River past New Orleans in a year.[469] The coal component was an additional three years.

The world's transition away from petroleum and, subsequently, away from coal was, of course, considerably more than an exercise in resourcefulness.

Those fuels were fundamental to civilization, and their exhaustion was like the failure of a building foundation pylon, something that could have meant catastrophe. Fortunately it didn't. The shortages arrived sufficiently slowly, at least when measured on human time scales, that people were able to deal with them incrementally and thus make it through the crisis in one piece. This is not to say they made it through peacefully. Who attacked whom and for what reason remain extremely confused today, presumably because of layers of denial over the many terrible things that occurred. You won't find agreement about it even within one country, much less across international borders. Written histories of the time always have such a strange mythic quality, as though they were partly made up, which of course they were. But it's best just to overlook these problems and move on, for it would have been too much to ask that humanity make it through this terrible time without any conflict.

Pondering how this great event played out really drives home to you that history is incomprehensibly long, just the way the ages of the rocks are, although of course not on the same scale—yet. Ordovician fish and dinosaurs rose, flowered, and became extinct, and the earth just ignored the whole thing and went on and on. Fuel from the ground got discovered, grew to dominate all economic activities of civilization, and then faded away, and history will no doubt forget the whole thing in just the same way as it moves on. You can even imagine the scurvy little footnote in textbooks four thousand years from now that students don't read because it's too boring. The end of fossil fuels seemed larger than life at the time, but it wasn't, for it was destined to disappear into the ancient past, forgotten, by virtue of being an event of the moment, a brief instant of geologic time.

It's cloudless this evening, and the sky is ablaze with stars, from old favorites like Rigel and Aldebaran to swarms of objects bright enough to have names but lacking them, on account of being too numerous. They're wonderfully magical, and also familiar old friends. The stars have, as far as we know, helped humans find their way across the earth and its oceans since they first walked upright. You're not just imagining it when the stars seem to make the night more primal still. They show who's boss even more persuasively than the cold does, being older than the earth in many cases, and as old as the universe itself in a few.[470] The perception of their permanence is an illusion, of course. The stars aren't friendly sprites in the sky but mighty nuclear furnaces. They're burning up their fuel, just as we are, and will one day exhaust this fuel and go out. A small fraction of them will die in glory, exploding with the brightness of ten billion suns, but the vast majority will just fade out ignominiously and go cold. Nothing lasts forever, either here on earth or in the heavens.

But the end of stars is a very different thing from the end of cheap carbon in the ground. The latter was an earthly problem, not a cosmic one, and it turned out to be surmountable. It was the coming of winter to people who had only known summer, for sure, but it wasn't anything more. The lights of civilization remained on for the simple reason that human beings are, in fact, rather resourceful. The lights are likely to remain on through future similar crises for the same reason. Exactly how history will play out thousands of years from now is anyone's guess, and we'd be foolish to speculate about it too much. However, it's a very safe bet that people of those times will have electricity, cars, and airplanes. These things are just too useful ever to relinquish, crises or no crises. The lights of civilization will eventually dim and perhaps go out when, in the far distant future, the night sky darkens, but that won't be for a very, very long time.

Meanwhile, the most terrible cosmic explosion of all will occur if I show up late again for dinner. It might be a good idea to stop worrying about the universe and hustle home.

ACKNOWLEDGMENTS

I wish to thank Nicholas Breznay, Nils Johan Engelsen, Albert Feng, Ching Hua Lee, Matt Potter, and Mike Rosen for helping error-correct the technical material, and also Sandy Fetter, Ted Geballe, and Peggy Laughlin for critical reading of the prose. Special thanks also go to Sissy and Ted Geballe for their steadfast support of scholarship over the years and for their encouragement of the young (and not-so-young) to tackle things that matter.

NOTES

CHAPTER 1: ARMCHAIR JOURNEY

1. There are few official statements, one way or the other, from either government or industry about the finiteness of geological energy supplies. It is also sometimes dismissed as a myth. See "Ending the Energy Stalemate," National Commission on Energy Policy, December 2004. However, there are a number of official publications surveying known and estimated geological fuel supplies, and finiteness is implicit in all of them. Three particularly important ones are "World Petroleum Assessment 2000," U.S. Geological Survey Digital Data Series, DDS-60; "BP Statistical Review of World Energy," British Petroleum, June 2010; "International Energy Annual (EIA)," U.S. Energy Information Administration, www.eia.gov /iea, last updated in 2006. Note that the latter is stable by virtue of having been discontinued (i.e., no longer continuously updated). The EIA and BP numbers largely agree except for petroleum, where they differ by 15% because of different assumptions about the density of crude oil.

2. This estimate for the exhaustion of fossil carbon is deliberately long. The U.S. Energy Information Administration's "International Energy Annual" quotes an exhaustion time for coal, the most plentiful fossil resource, of 9.09×10^{11} tons of reserve / 7.67×10^9 tons/year of consumption = 118 years (see note 1). The analogous calculation from the BP Statistical Review of World Energy is 8.26×10^{11} TOE of reserve / 3.28×10^9 TOE/year of consumption = 251 years (see note 1). The recent U.S. National Academy of Sciences report *Coal: Research and Development to Support National Energy Policy* (National Academies Press, 2007) revised this time down to one hundred years.

3. A. M. Jaffe, K. B. Medlock III and L. A. Smulcer, "U.S. Energy Policy FAQ: The U.S. Energy Mix, National Security, and the Myths of Energy Independence," James A. Baker III Institute for Public Policy, 1 Feb 08. This article somewhat understates

Director Jaffe's concerns about the oil future, which are considerable. See A. M. Jaffe, "We Won't Get There by Tinkering," Houston Chronicle, 24 Feb 08; A. M. Jaffe, "The Cowardly Giants," Newsweek, 26 Nov 07; T. Cabot, "The Oil Crisis: How Doomed Are We?," Esquire, August 2006, p. 50.

4. The mismatch between the energy problem's magnitude and government power is illustrated starkly by the U.S Energy Independence and Security Act of 2007. The law's most important provision was to raise annual domestic ethanol production from 9 billion gallons in 2008 to 36 billion gallons in 2022. The U.S. gasoline usage in 2022 is difficult to estimate accurately, but it is probably bracketed by the 2020 values of 9.73×10^6 bbl/d $\times$ 365 d/y $\times$ 42 gal/bbl = 1.49×10^{11} gals/y and 1.32×10^{11} gals/y in Table C4 of "Annual Energy Outlook 2008," U.S. Energy Information Administration, DOE/EIA-0383(2008), June 2008. (Note that these bounds do not include the E85 estimates of between 9.7 and 10.3 billion gallons per year.) Given that the energy content of ethanol is 2/3 that of gasoline, the fraction of the gasoline energy demand supplied by ethanol in 2020 would then have been $2/3 \times 3.6 \times 10^{10}$ gals / $(1.40 \times 1011$ gals$)$ = 0.17. However, this figure is much too high because (1) gasoline accounts for less than half of the 240 billion gallons expected (in "Annual Energy Outlook 2008") to be consumed by the United States in 2020, and (2) ethanol takes nearly as much energy to make as it yields. Estimates range from 1.6 to 0.6 joules of energy produced for each joule of energy expended. The smaller figure obviously corresponds to a net energy loss. Adopting the "consensus" yield of 1.3, and assuming that the extra energy is not supplied from external sources, such as nuclear reactors, the fraction of U.S. gasoline supplanted by ethanol in 2020 reduces to $(1.3 - 1)/1.3 \times 0.17$ = 0.039. The 2007 law also mandated that the bulk of the ethanol production increase come from "cellulosic technologies," which are estimated to have energy yields much greater than 1.3. These technologies do not yet exist at production level. Meanwhile, the recent recession threw three ethanol companies into bankruptcy.

See H. Shapouri, J. A. Duffield, and M. Wang, "The Energy Balance of Corn Ethanol: An Update," U.S. Department of Agriculture AER-814 (2002); A. E. Farrell et al., "Ethanol Can Contribute to Energy and Environmental Goals," Science **311**, 506 (2006); K. Bullis, "Will Cellulosic Ethanol Take Off?" Technology Review, 26 Feb 07; K. Galbraith, "In Gas-Powered World, Ethanol Stirs Up Complaints," New York Times, 26 Jul 08; K. Galbraith, "Economy Shifts, and the Ethanol Industry Reels," New York Times, 4 Nov 08; D. Diller and P. Brasher, "Plunging Profits Trump Ethanol's Previous Fears," Des Moines Register, 22 Dec 08.

5. Numerous sources give the average discharge of the Mississippi as 600,000 ft^3/ sec. According to the U.S. Energy Information Administration's "International Energy Annual (IEA)" (www.eia.giv/iea), the world consumed 85 million bbl of crude oil each day in 2006. The time it takes 85 million bbl of water to pass by New Orleans is 8.5×10^7 bbl $\times$ 5.6 ft^3/bbl / $(6 \times 10^5$ ft^3/sec $\times$ 60 sec/min$)$ = 13.2 minutes. The EIA also identifies the amount of energy consumed by oil, natural gas, coal,

and other (hydro, nuclear, wind, etc.) in 2006 to be 1.8, 1.1, 1.3, and 0.2 respectively, in units of 10^{20} joules/year. The equivalent Mississippi flow for the world's total energy consumption is thus 13 min × (1.8 + 1.1 + 1.3 + 0.2) / 1.8 = 32 minutes. The equivalent power is 4.4×10^{20} joules/y / (365 d/y × 24 h/d × 3,600 sec/h) = 1.40 × 10^{13} watts. These numbers cross-check with the BP Statistical Review of World Energy, which gives them as 1.6, 1.1, 1.3, and 0.6, assuming the standard conversion factor of 1 TOE (Tonne of Oil Equivalent) = 4.2×10^{10} joules. The disparity in oil component is due entirely to different assumptions about the mass per bbl. The number of bbl per year is the same in the two reports. The 2009 BP numbers are 1.7, 1.1, 1.4, and 0.6.

6. D. Yergin, *The Prize: The Epic Quest for Oil, Money, and Power* (Free Press, 1993); P. R. Osborn, *Operation Pike* (Greenwood Press, 2000); F. W. Engdahl, *A Century of War: Anglo-American Oil Politics and the New World Order* (Pluto Press, 2004).

CHAPTER 2: GEOLOGIC TIME

7. The value of 1 meter per year is approximate. Some specific values: Cairo 0.03, Riyadh 0.11, Athens 0.37, San Francisco 0.52, London 0.61, Seattle 0.88, Rio de Janiero 1.10, Shanghai 1.14, New York 1.20, Atlanta 1.26, Kinshasa 1.41, Bangkok 1.47, Tokyo 1.52, Mobile 1.53, Calcutta 1.63, Jakarta 1.82, Hong Kong 2.22, Singapore 2.27. Source: Global Historical Climatology Network, U.S. National Oceanic and Atmospheric Administration, as ported through www.worldclimate.com. Were this water to accumulate rather than draining away, it would create a layer around the earth 4,000 meters deep, the average depth of the oceans, in 4,000 years. However, only 0.7 of the earth's surface is ocean, so the time to fill up the oceans is only 0.7 × 4,000 years = 2,800 years. The time to deliver one earth volume of rain is $1/3 \times 6.37 \times 10^6$ meters / (1 meter/year) = 2.12×10^6 years.

8. The estimate of 2.2 billion years ago for the time when cyanobacteria created the air we now breathe comes mainly from chemical composition of ancient rocks. See L. Margulis and D. Sagan, *Microcosmos* (U. of California Press, 1997), and J. W. Schopf, "Solution to Darwin's Dilemma: Discovery of the Missing Precambrian Record of Life," Proc. Natl. Acad. Sci. **97**, 6947 (2000).

9. The prediction that CO_2 buildup will warm the earth is reliable, but the predictions of how warm the earth will get have large uncertainties. Small changes in the model assumptions, or simply computer errors, can make large changes in the final outcome. The present-day political controversy over whether global warming is real or "fraud" thus has a physical cause, namely the sensitivity of the temperature calculations to small details that people may or may not have gotten right. Obviously the world will eventually know whose predictions (if any) were correct, since it's in the process of conducting the experiment by doubling the CO_2 content of the atmosphere.

10. See the United Nations Framework Convention on Climate Change, as supplemented by the 1998 Kyoto Protocol. This treaty, which remains in force only through 2012, mandates reductions in greenhouse gas emissions, chiefly carbon dioxide, for the specific purpose of retarding global warming. The United States was alone among industrialized countries in refusing to ratify this treaty. Carbon dioxide emissions have continued to rise unabated since the treaty went into force. The most important increases have come from China. See G. Prins and S. Rayner, "Time to Ditch Kyoto," Nature **449**, 973 (2007); E. Rosenthal, "China Increases Lead as Biggest Carbon Dioxide Producer," New York Times, 14 Jun 08.

11. The political impossibility of keeping carbon in the ground in peacetime is indicated by a long, unhappy record of fuel strikes, partisan tax battles, and international treaty disappointments. See "Fuel Protests Build Across Europe," BBC News, 12 Sep 00; W. Tucker, "Carbon Limits, Yes; Energy Subsidies, No," Wall Street Journal, 28 Dec 08; J. van Loon, "Expectations Are Low for Climate Agreement, Germany Says," Bloomberg, 1 Dec 08.

12. The turnover rate of deep water in the oceans is measured by radiocarbon age differences of foraminifera shells deposited in the mud at different depths. The measurements are controversial in part because they require special technology to detect the extremely small quantities of ^{14}C involved. There are also questions about the integrity of the shell populations at great depths. See J. F. Adkins and C. Pasquero, "Deep Ocean Overturning—Then and Now," Science **306**, 1143 (2004); W. S. Broecker, "Deep Divisions," The Guardian, 18 Jun 00.

13. D. Archer *et al.*, "Atmospheric Lifetime of Fossil Fuel Carbon Dioxide," Annu. Rev. Earth Planet. Sci. **37**, 117 (2009). This paper's lowest model assumptions for the amount of carbon likely to go into the air match the BP Statistical Review of World Energy 2008 estimates for total carbon assets of the world: 1.09×10^{12} tonnes. The BP figure breaks down as 8.47×10^{11} tonnes from coal, $12/14 \times 1.68 \times 10^{11}$ tonnes $= 1.44 \times 10^{11}$ tonnes from oil, and $12/16 \times 0.67 \times 10^{-3}$ tonnes/m$^3 \times 1.77 \times 10^{14}m^3 = 0.89 \times 10^{11}$ tonnes for natural gas. The U.S. Geological Survey estimates are similar to BP's. The highest model assumption in Archer *et al.* is 5 times this amount. See also G.-K. Plattner *et al.*, "Feedback Mechanism and Sensitivities of Ocean Carbon Uptake Under Global Warming," Tellus **35**, 564 (2001).

14. P. A. Colinvaux, *Why All Big Fierce Animals Are Rare* (Princeton U. Press, 1979).

15. S. Minobe, "A 50–70 Year Climatic Oscillation over the North Pacific and North America," Geophys. Res. Lett. **24**, 683 (1997); K. E. Trenberth and J. M. Caron, "The Southern Oscillation Revisited: Sea Level Pressures, Surface Temperatures, and Precipitation," J. Climate **13**, 4358 (2000); M. S. Barlow, S. Nigam, and E. H. Berbey, "ENSO Pacific Decadal Variability and U.S. Summertime Precipitation, Drought, Precipitation and Stream Flow," J. Climate **14**, 2105 (2001); E. DiLorenzo *et al.*, "North Pacific Gyre Oscillation Links Ocean Climate and Ecosystem Change," Geophys. Res. Lett. **35**, 6 (2008); M. E. Linkin and S. Nigam, "The

North Pacific Oscillation-West Pacific Teleconnection Pattern: Mature-Phase Structure and Winter Impacts," J. Climate **21**, 1979 (2008).

16. E. O. Wilson, "Is Humanity Suicidal?," New York Times Magazine, 30 May 93; R. E. Leakey and R. Lewin, *The Sixth Extinction* (Anchor, 1996); N. Meyers and A. H. Knoll, "The Biotic Crisis and the Future of Evolution," Proc. Natl. Acad. Sci. **98**, 5389 (2001); A. D. Barnosky *et al.*, "Has the Earth's Sixth Mass Extinction Event Already Arrived?," Nature **471**, 51 (2011).

17. Walter Alvarez's comet impact theory of dinosaur extinction is supported by trace amounts of iridium metal at the cretaceous-tertiary boundary. There are also less violent theories of how the dinosaurs died. See W. Alvarez, *T. Rex and the Crater of Doom* (Princeton, 2008); S. E. Peters, "Environmental Determinants of Extinction Selectivity in the Fossil Record," Nature **454**, 626 (2008). An asteroid ten kilometers in diameter made of rock (specific gravity 3) and impacting at earth's escape velocity (1.1×10^4 m/sec) would have explosive energy $4\pi/3 \times (5000 \text{ m})^3 \times 3000 \text{ kg/m}^3 \times 9.8$ m/sec$^2 \times 6.37 \times 10^6$ m $= 9.81 \times 10^{22}$ joules. Since a ton of TNT is 4.184×10^9 joules, this asteroid energy is 2.34×10^{13} tons of TNT or 0.23 million 100-megaton warheads.

18. R. Stone, "A World Without Corals," Science **316**, 678 (2007); K. E. Carpenter *et al.*, "One-Third of Reef-Building Corals Face Elevated Extinction Risk from Climate Change and Local Impacts," Science **321**, 560 (2008); B. Walsh, "Coral Reefs Face Extinction," Time, 11 Jul 08.

19. J. M. Kiesecker, A. R. Blaustein, and L. K. Belden, "Complex Causes of Amphibian Population Declines," Nature **410**, 639 (2001); J. A. Pounds *et al.*, "Widespread Amphibian Extinctions from Epidemic Disease Driven by Global Warming," Nature **439**, 161 (2006); I. Sample, "Over Half of Europe's Amphibians Face Extinction by 2050," The Guardian, 26 Sep 08.

20. P. R. Ehrlich and A. H. Ehrlich, *One with Nineveh: Politics, Consumption and the Human Future* (Island Press, 2004).

21. Carbon dioxide presently constitutes 0.0386% by volume of air. This corresponds to $[CO_2]_{air} = 3.86 \times 10^{-4} \times 10^{-3}$ m^3/liter $\times 1.013 \times 10^5$ pascals / (8.314 joules $°K^{-1} \times 293°K$) $= 1.60 \times 10^{-5}$ moles/liter. With the carbon dioxide concentration of the water denoted by $[CO_2]$, the mass-action equilibrium equations are (1) $[CO_2]/[CO_2]_{air} = 0.83$, (2) $[HCO_3^-][H^+]/[CO_2] = 4.25 \times 10^{-7}$ moles/liter, (3) $[OH^-][H^+] = 10^{-14}$ (moles/liter)^2, and (4) $[HCO_3^-] + [OH^-] = [H^+]$. The solution is $[CO_2] = 1.33 \times 10^{-5}$ moles/liter, $[HCO_3^-] = 2.37 \times 10^{-6}$ moles/liter, $[H^+] = 2.38 \times 10^{-6}$ moles/liter, and $[OH^-] = 4.21 \times 10^{-9}$ moles/liter. We then obtain pH $= -\log_{10}[H^+] = 5.62$. The measured pH of distilled water exposed to air lies between 5.5 and 5.8, the uncertainty being due to slight variations in the local CO_2 concentration of the air. See B. M. Mahan and R. J. Meyers, *University Chemistry* (Benjamin-Cummings, 1987); E.-G. Beck, "50 Years of Continuous Measurement of CO_2 on Mauna Loa," Energy and Environment **19**, 1017 (2008).

22. Calcium carbonate in contact with distilled water is described by the first three equations above plus (4) $[CO_3^{2-}][H^+]/[HCO_3^-] = 5.61 \times 10^{-11}$ moles/liter, (5) $[CO_3^{2-}]$

$[Ca^{2+}] = 4.47 \times 10^{-9}$ (moles/liter)^2, and (6) $2\,[CO_3^{2-}] + [HCO_3^-] + [OH^-] = 2\,[Ca^{2+}] + [H^+]$. The solution is $[CO_2] = 1.33 \times 10^{-5}$ moles/liter, $[HCO_3^-] = 9.58 \times 10^{-4}$ moles/liter, $[CO_3^{2-}] = 9.13 \times 10^{-6}$ moles/liter, $[Ca^{2+}] = 4.90 \times 10^{-4}$ moles/liter, $[H^+] = 5.89 \times 10^{-9}$ moles/liter, and $[OH^-] = 1.70 \times 10^{-6}$ moles/liter. We then obtain pH $= -\log_{10}[H^+] = 8.23$. The measured pH of seawater ranges from 7.5 to 8.4, the uncertainty being due chiefly to influx of freshwater from rivers. The water's total carbon concentration compared to that of air is $([CO_2] + [HCO_3^-] + [CO_3^{2-}])/[CO_2]_{air} = 61.3$. The fraction of the water's carbon that came from the calcium carbonate is $[Ca^{2+}] / ([CO_2] + [HCO_3^-] + [CO_3^{2-}]) = 0.50$. See J. N. Butler, *Carbon Dioxide Equilibria and Their Applications* (CRC Press, 1991).

23. The atmosphere has an effective thickness (exponential decay depth) of 8,000 meters. Were the oceans filled with distilled water in contact with limestone, per the previous example, they would hold more carbon than the atmosphere by the factor $61.3 \times 0.7 \times 4{,}000$ meters / (8,000 meters) $= 21.5$. The total mass of carbon would be $0.7 \times 4\pi \times (6.37 \times 10^6 \text{ m})^2 \times 4{,}000 \text{ m} \times 10^3 \text{ liters/m}^3 \times 9.80 \times 10^{-4}$ moles/liter $\times 12.0 \times 10^{-6}$ tonnes/mole $= 1.68 \times 10^{13}$ tonnes. The carbon content of actual seawater is slightly more than twice this idealized value.

24. According to a recent USA Today/Gallup Poll, two-thirds of Americans say that creationism is "definitely or probably true." Creationism holds that that the earth and its life are less than 10,000 years old. See J. Lawrence, "Poll Shows Belief in Evolution, Creationism," USA Today, 7 Jun 07. Creationist objections against determinations of geologic time are spelled out in J. Woodmorappe, *Mythology of Modern Dating Methods* (Inst. for Creation Research, 1999).

25. The site in question is San Gregorio Beach, about forty miles south of San Francisco, California. Pictures of the cliffs are posted at www.shallowsky.com.

26. The sea is presently rising at about two millimeters per year, perhaps a bit more. There is some indirect evidence that it was rising less rapidly just before the industrial revolution, but there were no systematic measurements at that time. There is also evidence that sea level may have risen a few meters since the end of the ice age, although this assessment is complicated by the continuing uplift of the continents in response to the relief of the ice burden. The sea rose extremely rapidly when the ice age ended. See N. C. Flemming, "Multiple Regression Analysis of Earth Movements and Eustatic Sea-Level Changes in the United Kingdom in the Last 9000 Years," Proc. Geol. Assn. **93**, 113 (1982); M. F. Meier *et al.*, "Glaciers Dominate Eustatic Sea-Level Rise in the 21st Century," Science **317**, 1064 (2007).

27. These particular rocks, identified by the U.S. Geological Survey as the Tahana Member of the Purisima Formation, are dated as Pliocene (1.8–5.33 million years ago) or Upper Pliocene (5.33–20.03 million years ago). They contain a layer of volcanic ash tentatively identified as the Ishi tuff, produced 2.6 million years ago. See E. E. Brabb, R. W. Graymere, and D. L. Jones, "Geology of the Onshore Part of San Mateo County, California: A Digital Database," U.S. Geological Survey, Open-File Report 98-137, 1998.

28. M. J. Aitken, "Archaeological Dating Using Physical Phenomena," Rep. Prog. Phys. **62**, 1333 (1999); G. B. Dalrymple, *The Age of the Earth* (Stanford U. Press, 1991).

29. A prototypical case of potassium-argon dating is G. B. Dalrymple and K. Hirooka, "Variation of Potassium, Argon, and Calculated Age in Late Cenozoic Basalt," J. Geophys. Res. **70**, 5291 (1965). Rock samples weighing roughly 10 grams were melted to release trapped gases. The amount of argon in the gas was measured by dilution mass spectrometry to be 1.3×10^{-11} moles/gram. The samples were then digested with acid and found by flame photometry to contain 2.6% by weight of K_2O. This corresponds to $2 \times 2.6 \times 10^{-2} / (2 \times 39.1 \text{ grams/mole} + 16 \text{ grams/mole}) = 5.52 \times 10^{-4}$ moles/gram of potassium. The age of the rock is then 1.44×10^{14} years $\times 1.3 \times 10^{-11}$ moles/gram $/ (5.52 \times 10^{-4}$ moles/gram$) = 3.39 \times 10^{6}$ years. Of the K atoms in the environment, 0.0117% are the radioactive isotope ^{40}K. These β-decay with a half-life of 1.248 billion years. Of these decays, 10.72% produce ^{40}Ar. (The rest produce ^{40}Ca.) The time coefficient is thus 1.248×10^9 years $/ (\ln(2) \times 0.1072 \times 1.17 \times 10^{-14}) = 1.44 \times 10^{14}$ years. Dilution mass spectrometry works by injecting a known amount (in this case 3×10^{-10} moles) of ^{38}Ar into the gas sample and measuring $[^{38}Ar]/[^{40}Ar]$.

30. C. Darwin, *On the Origin of Species by Means of Natural Selection, or the Preservation of Favored Races in the Struggle for Life* (John Murray, 1859). This first edition was followed by subsequent ones in 1860, 1861, 1866, 1869, and 1872, the sixth being the version commonly used today. Darwin argued in the first edition that the Weald was a dome-like uplift that had eroded away, exposing the older rock at the center. He estimated its age by modeling the erosion as seaside cliff retreat. Assuming that 500-foot cliffs retreat at 1 inch/century, that the cliff in question was 1,100 feet high (the depth of the eroded chalk), and that the retreat distance was 22 miles (the width of the Weald), he obtained 1,100 ft $\times$ 22 miles $\times$ 63,360 in/mile $/ (500 \text{ ft} \times 0.01 \text{ in/year}) = 3.07 \times 10^8$ years.

31. There are no volcanic rocks in the Weald, unfortunately, so it must be dated indirectly through volcanic intrusions into similar rocks elsewhere in the world. The lower greensand at its center is usually dated as Aptian (112–125 million years ago). See P. Skelton *et al.*, *The Cretaceous World* (Cambridge U. Press, 2003).

32. D. R. Dean, *Gideon Mantell and the Discovery of Dinosaurs* (Cambridge U. Press, 1999); G. A. Mantell, "Notice on the Iguanodon, a Newly Discovered Fossil Reptile, from the Sandstone of Tilgate Forest, in Sussex," Phil. Trans. Roy. Soc. **115**, 179 (1825).

33. The remarkable structure of Great Britain is discussed in most elementary geology textbooks. A particularly nice set of geological maps from H. M. Woodward, *Stanford's Geological Atlas of Great Britain and Ireland* (Edward Stanford, 1904) is available at the Internet site of Ian West, School of Ocean and Earth Sciences, University of Southampton (www.soton.ac.uk/~imw/Geology-Britain.htm).

34. The literature of radiometric dating is too large to cite comprehensively. Here are some representative examples: M. E. Smith *et al.*, "High-Resolution Calibration

of Eocene Strata: ^{40}Ar/^{39}Ar Geochronology of Biotite in the Green River Formation," Geology **34**, 393 (2006) [50 million years]; P. R. Renne *et al.*, "The Age of Paraná Flood Volcanism, Rifting of Gondwanaland, and the Jurassic-Cretaceous Boundary," Science **258**, 975 (1992) [110 million years]; J. C. Brannon *et al.*, "Th-Pb and U-Pb Dating of Ore-Stage Calcite and Paleozoic Fluid Flow," Science **271**, 491 (1996) [251 million years]; E. Landing *et al.*, "Cambrian-Ordovician Boundary Age and Duration of the Lowest Ordovician Temadoc Series Based on U-Pb Zircon Dates from Avalonian Wales," Geol. Mag. **137**, 485 (2000) [490 million years]; P. W. G. Tanner and J. A. Evans, "Late Precambrian U-Pb Titanate Age for Peak Regional Metamorphism and Deformation (Knoydartian Orogony) in the Western Moine, Scotland," J. Geol. Soc. **160**, 555 (2003) [707 million years].

35. The literature of precambrian radiometric dating is likewise too large to cite comprehensively. Here are some representative examples: O. van Breeman, M. Aftalion, and J. H. Allaart, "Isotropic and Geochronologic Studies on Granites from the Ketilidian Mobile Belt of South Greenland," Geol. Soc. Am. Bull. **85**, 403 (1974) [2 billion years]; S. Moorbath, "First Direct Radiometric Dating of Archaean Stromatolite Limestone," Nature **326**, 865 (1987) [3 billion years]; S. A. Bowring and I. S. Williams, "Priscoan Orthogneisses from Northwestern Canada," Contrib. Minerol. Petrol. **134**, 3 (1999) [4 billion years]; S. A. Wilde *et al.*, "Evidence from Detrital Zircons for the Existence of Continental Crust and Oceans on the Earth 4.4 Gyr Ago," Nature **409**, 175 (2001) [4.4 billion years]. There are still questions about the legitimacy of dates from the first billion years of geologic time.

36. C. Patterson, "Age of Meteorites and the Earth," Geochim. Cosmochim. Acta **10**, 230 (1956); E. Anders, "Meteorite Ages," Rev. Mod. Phys. **34**, 287 (1962); R. Hutchinson, "Meteorite Ages," Nature **308**, 500 (1984); J. J. Papike, G. Ryder, and C. K. Shearer, "Lunar Samples," Rev. Minerol. Geochem. **36**, 1 (1998); A. Nemchin *et al.*, "Timing of Crystallization of the Lunar Magma Ocean Constrained by the Oldest Zircon," Nature Geosci. **2**, 133 (2009).

37. The sedimentary rocks below the Gulf of Mexico, the Persian Gulf, and the North Sea are all more than 12 kilometers thick. See D. E. Bird, K. Burke, S. A. Hall, and J. F. Casey, "Gulf of Mexico Tectonic History: Hotspot Tracks, Crustal Boundaries, and Early Salt Distribution," Am. Assoc. Petroleum Geol. Bull. **89**, 311 (2005); A. I. Konyukov and B. Maleki, "The Persian Gulf Basin: Geological History, Sedimentary Formations and Petroleum," Lithology and Mineral Resources **41**, 344 (2006); G. G. Leckie, "Lithology and Subsidence in the North Sea," Phil. Trans. Roy. Soc. A **305**, 85 (1982). The Anadarko Basin in Oklahoma, the location of several very deep natural gas wells, is 40,000 feet thick in places. It's the location of the famous Bertha Rogers well, which was drilled through sedimentary rocks to 31,000 feet, where it stopped because of striking liquid sulfur. Other famous deep wells include the Burlington Big Horn 6-67 in Fremont County, Wyoming, which found natural gas at 26,000 feet in Madison (i.e., Carboniferous) limestone, and the Chevron Knotty Head strike at 30,589 feet in the Gulf of Mexico.

38. D. L. Baars, *The Colorado Plateau* (U. New Mexico Press, 2000).

39. The literature of thick sediments is again too huge to site comprehensively. Here are some representative examples: J. T. Whetten, "Wairakite from Low-Grade Metamorphic Rocks on St. Croix, U.S. Virgin Islands," Am. Mineralogist **50**, 752 (1965); M. Kay, "North Atlantic Continental Drift," Proc. Am. Phil. Soc. **112**, 321 (1968); K. Kamada, "Coarse Clastic Sedimentation in the Triassic Offshore Sequence of the Southeastern Kitakami Mountains, Japan," in *Sedimentary Facies in the Active Plate Margin*, A. Taira and F. Masuda, eds. (Terra Sci., 1998), p. 365; P. G. Eriksson and W. Alterman, "An Overview of the Geology of the Transvaal Supergroup Dolomites," Environ. Geol. **36**, 179 (1998); E. J. Milani and P. V. Zalán, "An Outline of the Geology and Petroleum Systems of the Paleozoic Interior Basins of South America," Episode **22**, 199 (1999); M. B. Miller and L. A. Wright, *The Geology of Death Valley* (Kendall/Hunt, 2004).

40. The stratigraphic determination of ancient sea levels is a laborious process involving judgments that are not always transparent because they're partly based on proprietary oil drilling logs. See B. U. Haq, J. Harnenbob, and P. R. Vail, "Chronology of Fluctuating Sea Levels Since the Triassic," Science **235**, 1156 (1987); B. U. Haq and S. R. Schutter, "A Chronology of Paleozoic Sea-Level Changes," Science **322**, 64 (2008).

41. Theoretically compensating for crustal deformation, a process called backstripping, reduces the long-term sea level changes by roughly a factor of 2. See K. G. Miller *et al.*, "The Phanerozoic Record of Global Sea-Level Change," Science **310**, 1293 (2005).

42. C. Emiliani, "Interglacial High Sea Levels and the Control of Greenland Ice by the Precession of the Equinoxes," Science **154**, 851 (1966); W. L. Prell *et al.*, "Graphic Correlation of Oxygen Isotope Stratigraphy: Application to the Late Quaternary," Paleoceanography **1**, 137 (1986); D. B. Karner *et al.*, "Constructing a Stacked Benthic $\delta^{18}O$ Record," Paleoceanography **17**, 1030 (2002); L. E. Lisiecki and M. E. Ramo, "A Pliocene-Pleistocene Stack of 57 Globally Distributed Benthic $\delta^{18}O$ Records," Paleoceanography **20**, PA1003, PA 2007 (2005). The $\delta^{18}O$ assessments of sea level changes further back than 1 million years are problematic because they lack ice core time fiducials.

43. The polar ice sheets are depleted in ^{18}O with respect to seawater for complex reasons having to do with evaporation and precipitation kinetics. The depletion is described roughly by Dansgaard's formula $\delta[^{18}O]_{ice}/[^{18}O]_{ice} = -1.4 \times 10^{-2} + 6.7 \times 10^{-4}$ T, where T is the annual mean temperature of the deposition site in degrees centigrade. Local measurements of ^{18}O content of polar ice cores thus provide a thermometric record of the ice ages—although the reliability of the calibration is routinely questioned. See W. Dansgaard, "Stable Isotopes in Precipitation," Tellus **16**, 426 (1964); M. P. Lawson, K. C. Kuivinen, and R. C. Balling Jr., "Analysis of the Climatic Signal in the South Dome, Greenland Ice Core," Climatic Change **4**, 375 (1982); E. A. Boyle, "Cool Tropical Temperatures Shift the Global $\delta^{18}O$-T Relationship: An

Explanation for the Ice Core δ^{18}O-Borehole Thermometry Conflict?," Geophys. Res. Lett. **24**, 273 (1997). As the polar ice sheets grow, the ^{18}O concentration in the sea becomes enhanced according to $\delta[^{18}O]_{sea}/[^{18}O]_{sea} = \delta[^{18}O]_{ice}/[^{18}O]_{ice} \times \delta h/(4{,}000$ meters $- \delta h)$, where δh is the change in sea level. Substituting a "typical" value of T $= -31°$C into Dansgaard's formula, we obtain $\delta[^{18}O]_{sea}/[^{18}O]_{sea} = -\delta h \ / \ 115{,}000$ meters. The enhancement is then recorded in the carbonate shells that foraminifera leave behind in the ocean sediments. The observed values of $\delta[^{18}O]$ tend to be about 30% higher than this choice of T predicts. See N. J. Shackleton, "Oxygen Isotopes, Ice Volume and Sea Level," Quat. Sci. Rev. **6**, 183 (1987).

44. Geologists knew long before the invention of ^{18}O technologies that the oldest of the four major glaciations (Nebraskan, in North American terminology) was an oversimplification. See R. F. Flint, *Glacial and Pleistocene Geology* (Wiley, 1957).

45. See, for example, J. Chapell *et al.*, "Reconciliation of Late Quaternary Sea Levels Derived from Coral Terraces at Huon Peninsula with Deep Sea Oxygen Isotope Records," Earth Planet. Sci. **141**, 227 (1996).

46. Evidence in the rocks reveals that most of Canada was covered during the last major glaciation and that the glaciers were typically 10,000 feet thick. The area of Canada is roughly nine million square kilometers. Assuming that the glaciated area was two Canadas and that the glaciers were 3,000 meters thick, the sea level depression was $2 \times 9.0 \times 10^{12}$ meters$^2 \times 3{,}000$ meters $/ \ (0.7 \times 4\pi \times (6.37 \times 10^6$ meters$)^2) = 151$ meters.

47. J. R. Petit *et al.*, "Climate and Atmospheric History of the Past 420,000 Years from the Vostok Ice Core, Antarctica," Nature **399**, 429 (1999). This paper shows a strong correlation between $\delta^{18}O_{atm}$ (changes in the ^{18}O content of O_2 trapped in air bubbles) and the precession of earth's polar axis, which affects climate indirectly through earth's orbital ellipticity. The moon and sun torque the earth through its equatorial bulge, which is determined by centrifugal force and isostasy. Were only the moon involved, the precession period would be $2/3 \times (5.97 \times 10^{24}$ kg $/ \ 7.35 \times 10^{22}$ kg$) \times (3.84 \times 10^8$ m $/ \ 6.38 \times 10^6$ m$)^3 \times 1$ day $/ \ (\cos 23° \times 365$ days/year$) = 3.51 \times 10^4$ years. The effect of the sun is mainly to reduce this by the factor $1 + (1.99 \times 10^{30}$ kg $/ \ 7.35 \times 10^{22}$ kg$) \times (3.84 \times 10^8$ m $/ \ 1.49 \times 10^{11}$ m$)^3 = 1.46$, thus making the period 2.40×10^4 years. The measured precession period is 7.4% greater than this value (25,772 years), due to details omitted from the calculation. See also EPICA Members, "Eight Glacial Cycles from an Antarctic Core," Nature **429**, 623 (2004); D. Lüthi *et al.*, "High-Resolution Carbon Dioxide Concentration Record 650,000–800,000 Years Before Present," Nature **453**, 379 (2008). The glaciation episodes show up clearly in the ice cores as sawtooth patterns in the carbon dioxide, deuterium, and water δ^{18}O signals.

48. See, for example, D. Garner and J. L. Forsythe, "A Radiocarbon Date from the Hartwell Moraine, Warron County, Ohio," Ohio J. Sci. **65**, 94 (1965); D. J. Esterbrook, "Radiocarbon Chronology of Late Pleistocene Deposits in Northwest Washington," Science **152**, 764 (1966); S. C. Porter and T. W. Swanson, "Radiocar-

bon Age Constraints on Rates of Advance and Retreat of the Puget Lobe of the Cordilleran Ice Sheet During the Last Glaciation," Quat. Res. **50**, 205 (1998).

49. The typical melt rate at the end of the last ice episode was slightly larger, about 1.45 cm/year, and spiked to 4 cm/year during a 500-year interval called Meltwater Pulse 1A. The flow during this interval was $0.7 \times 4\pi \times (6.37 \times 10^6 \text{ meters})^2 \times 0.04$ meters /year = 1.43×10^{13} meters^3/year. Since ice's latent heat of melting is 3.34×10^8 joules/meter^3, the heat required was 4.78×10^{21} joules/year, 10.8 times the present-day energy budget of civilization (see note 5). Various sources identify the present flow of the Amazon to be about 7.5×10^6 ft^3/sec $\times 3.15 \times 10^7$ sec/year / $(3.281 \text{ ft/m})^3$ = 6.68 $\times 10^{12}$ m^3/year. The Amazon presently accounts for 20% of all freshwater flow to the sea. The latter is thus about 3.34×10^{13} m^3/year. It is only one-fifth of the $0.3 \times 4\pi$ $\times (6.37 \times 10^6 \text{ m})^2 \times 1\text{m/year} = 1.53 \times 10^{14}$ m^3/year of rain reported to fall on land (see note 7). Most of the disparity is due to evaporation. (The U.S. Geological Survey estimates that roughly 70% of the precipitation that falls on the continental United States evaporates.) The remaining disparity comes from estimation and measurement error. See A. J. Weaver *et al.*, "Meltwater Pulse 1A from Antarctica as a Trigger of the Bolling-Allerold Warm Interval," Science **299**, 1709 (2003).

50. There is an enormous literature on paleoclimatology. Good introductions may be found in B. T. Huber, K. G. MacLeod, and S. L. Wing, *Warm Climates in Earth History* (Cambridge U. Press, 2000); B. Saltzman, *Dynamical Paleoclimatology, Vol. 80: Generalized Theory of Global Climate Change* (Academic Press, 2001); H. C. Jenkyns, "Evidence for Rapid Climate Change in the Mesozoic-Palaeogene Greenhouse World," Phil. Trans. Roy. Soc. A **361**, 1885 (2003); and M. Williams *et al.*, eds., *Deep-Time Perspectives on Climate Change: Marrying the Signal from Computer Models and Biological Proxies* (Geological Soc. London, 2008).

51. The desiccation and subsequent refilling of the Mediterranean about six million years ago is recorded in undersea sediments as salt layers. See Duggen *et al.*, "Deep Roots of the Messinian Salinity Crisis," Nature **422**, 602 (2003); K. J. Hsu, *The Mediterranean Was a Desert: The Voyage of the Glomar Challenger* (Princeton U. Press, 1987).

52. A. B. Herman and R. A. Spicer, "Palaeobotanical Evidence for a Warm Cretaceous Arctic Ocean," Nature **380**, 330 (1996); B. T. Huber, "Tropical Paradise at the Cretaceous Poles?," Science **282**, 2199 (1998); J. A. Tarduno *et al.*, "Evidence for Extreme Climatic Warmth from Late Cretaceous Arctic Vertebrates," Science **282**, 2441 (1998); M. Friedman, J. A. Tarduno, and D. B. Brinkman, "Fossil Fishes from the High Canadian Arctic: Further Palaeobiological Evidence for Extreme Climatic Warmth During the Late Cretaceous (Turonian-Coniacian)," Cretaceous Res. **24**, 615 (2003).

53. J. M. Schopf, "Petrified Peat from a Permian Coal Bed in Antarctica," Science **169**, 274 (1970).

54. J. D. Hayes, J. Imbrie, and N. J. Shackleton, "Variations in the Earth's Orbit: Pacemaker of the Ice Ages," Science **194**, 1121 (1976); J. Imbrie and K. P. Imbrie,

Ice Ages: Solving the Mystery (Harvard U. Press, 1986); J. Imbrie, A. C. Mix, and D. G. Martinson, "Milankovitch Theory Viewed from Devils Hole," Nature **363**, 531 (1993); R. A. Muller and G. J. MacDonald, "Glacial Cycles and Astronomical Forcing," Science **277**, 215 (1997); D. B. Karner and R. A. Muller, "A Causality Problem for Milankovitch," Science **288**, 2143 (2000).

55. W. S. Boercker, "Was a Change in Thermohaline Circulation Responsible for the Little Ice Age?," Proc. Natl. Acad. Sci. **97**, 1339 (2000).

56. J. P. Kennett *et al.*, "Carbon Isotopic Evidence for Methane Hydrate Instability During Quaternary Interglacials," Science **288**, 128 (2000); T. de Garidel-Thoran *et al.*, "Evidence for Large Methane Releases to the Atmosphere from Deep-Sea Gas-Hydrate Dissociation During the Last Glacial Episode," Proc. Natl. Acad. Sci. **101**, 9187 (2004); T. Sowers, "Late Quaternary Atmospheric CH_4 Isotope Record Suggests Marine Clathrates Are Stable," Science **311**, 838 (2006); M. Kennedy, D. Mrofka, and C. von der Borch, "Snowball Earth Termination by Destabilization of Equatorial Permafrost Methane Clathrate," Nature **453**, 642 (2008); R. Boswell, "Is Gas Hydrate Energy within Reach?," Science **325**, 957 (2009).

57. J. A. Curry, J. L. Schramm, and E. E. Ebert, "Sea Ice-Albedo Climate Feedback Mechanism," J. Climate **8**, 240 (1995).

58. M. E. Raymo and W. F. Ruddiman, "Tectonic Forcing of Late Cenozoic Climate," Nature **359**, 117 (1992).

59. D. W. Patten, *The Biblical Flood and the Ice Epoch* (Pacific Meridian Press, 1966).

60. Were the earth a blackbody, i.e., a perfect absorber and emitter of light, and also forced by the winds to have a uniform surface temperature, this temperature would be fixed by the distance, size, and temperature of the sun to be $(0.5 \times 6.96 \times 10^8$ m / 1.50×10^{11} m$)^{1/2} \times 5778°$K = $278.3°$K, or $5.2°$C. The accepted average temperature of the earth at present is about $15°$C. The ten-degree difference is due to a number of factors left out of the calculation, including the greenhouse effect. See Erbe Science Team, "First Data from the Earth Radiation Budget Experiment," Bull. Am. Meteorol. Soc. **67**, 818 (1986); V. Ramanathan *et al.*, "Cloud-Radiation Forcing and Climate: Results from the Earth Radiation Budget Experiment," Science **243**, 57 (1989).

61. J. Zachos *et al.*, "Trends, Rhythms and Aberrations in Global Climate 65 Ma to Present," Science **292**, 686 (2001).

CHAPTER 3: JUNGLE LAW

62. The great majority of people in developing countries are not, of course, driving expensive cars. They just wish they were driving expensive cars. See E. Rosenthal, "Rich-Poor Divide Still Stalls Climate Accord," New York Times, 10 Apr 09; K. Bradsher, "Paying in Pollution for Energy Hunger," New York Times, 9 Jan 07.

63. F. Castro, "The Law of the Jungle," Counterpunch, 13 Oct 08. This item from Fidel Castro's blog is particularly popular and posted in many places on the Internet.

64. The consolidation of the oil industry in the late nineteenth century was all about cutthroat pricing to drive out competition. See I. M. Tarbell, *The History of the Standard Oil Company* (Dover, 2003).

65. The shale oil industry has come and gone repeatedly, killed off each time by the low cost of conventional crude. Oil shale is a sedimentary rock containing kerogen, the decomposed remains of algae and vascular plants often identified as the precursor to petroleum, and from which synthetic petroleum can be obtained by distillation. The world's oil shale deposits are estimated at 2.8 trillion barrels of oil equivalent, or about the same as the world's proved crude oil reserves. See A. Andrews, "Oil Shale: History, Incentives and Policy," CRS Report for Congress RL33359, 13 Apr 06; J. R. Dyni, "Geology and Resources of Some World Oil-Shale Deposits," U.S. Geological Survey Scientific Investigations Report 2005-5294, June 2006; D. R. Steuart, "The Shale Oil Industry of Scotland," Economic Geology **3**, 573 (1908); V. C. Alderson, *The Oil Shale Industry* (Frederick A. Stokes, 1920).

66. The estimated oil sands reserves in Canada are about 1.7×10^{12} bbl, an amount comparable to all the oil in Saudi Arabia. Venezuela has similar reserves. See "Alberta's Energy Reserves 2007 and Supply/Demand Outlook 2008–2017," Energy Resources Conservation Board Report ST98-2008, June 2008. Canadian oil sand mining has recently come under attack from environmental groups. See J. Simpson, "Alberta's Tar Sands Are Soaking Up Too Much Water," Globe and Mail, 4 Jul 06; A. Nikiforuk, "Canada's Highway to Hell," Onearth, 1 Sep 07; A. Edemariam, "Mud, Sweat and Tears," The Guardian, 30 Oct 07; M. Griffiths, A. Taylor and D. Woynillowicz, "Troubled Waters, Troubling Trends: Technology and Policy Options to Reduce Water Use in Oil and Oil Sands Development in Alberta," The Pembina Institute, May 2006.

67. The coal industry is notorious for its constant cost retrenchments in response to paralyzing strikes. The history of coal is intertwined with labor issues and is thus particularly difficult to discuss objectively. Representative publications: J. Grossman, "The Coal Strike of 1902—Turning Point in U.S. Policy," Monthly Labor Review **98**, 21 (1975); G. S. McGovern and L. F. Guttridge, *The Great Coalfield War* (Houghton Mifflin, 1972); R. A. Brisbin Jr., *A Strike Like No Other Strike: Law and Resistance During the Pittston Coal Strike of 1989–1990* (Johns Hopkins U. Press, 2002); H. B. Lee, *Bloodletting in Appalachia: The Story of West Virginia's Four Major Mine Wars and Other Thrilling Incidents on Its Coal Fields* (West Virginia U. Press, 1969); J. W. Hevener, *Which Side Are You On?: The Harlan County Coal Miners 1931–39* (U. Illinois Press, 1989).

68. The historically important 1984 Miners' Strike is extraordinarily well documented in a twenty-four-article BBC Online 20th anniversary retrospective. See "Watching the Pits Disappear," BBC News, 5 Mar 04; P. Hetherington, "Hidden Legacy of Pit Closures," The Guardian, 4 Mar 05.

69. A good overview of the California energy crisis was published as a three-part feature in the *San Francisco Chronicle*: M. Martin and L. Gledhill, "Paying the

State's Price of Power," 23 Dec 01; C. Said, "Deregulation Folly Is No Laughing Matter," 24 Dec 01; B. Tansey, "Deregulation (Sort of) Lives on in State," 25 Dec 01.

70. "The Electric Utility Industry Restructuring Act," California Assembly Bill No. 1890 (Statutes of 1996, Chapter 854, Brulte). Senator Steve Peace's role in drafting the legislation is described in C. Berthelsen, "Genesis of State's Energy Fiasco," San Francisco Chronicle, 31 Dec 00.

71. D. Lazarus, "Summer Ushered in a Power Crisis That Promises Only to Get Worse," San Francisco Chronicle, 29 Dec 00; D. Morain and N. Rivera Brooks, "Rolling Blackouts Hit Southland for First Time as Production Falls," Los Angeles Times, 20 Mar 01.

72. D. Lazarus, "PG&E Files for Bankruptcy," San Francisco Chronicle, 7 Apr 01.

73. G. Skelton, "Davis Says Bad Guys Went Thataway, to Texas," Los Angeles Times, 21 May 01; G. Davis, "More Than California's Problem," Washington Post, 16 May 01.

74. The companies included Dynegy and El Paso (both headquartered in Houston), Calpine, Mirant, and Williams. See L. M. Holson and R. A. Oppel Jr., "Long-Term Power Deals Scrutinized in California," New York Times, 16 Jun 01.

75. The concept of deregulating the U.S. electric power industry at the wholesale level, including the creation of the Independent System Operator (ISO), is detailed in Federal Energy Regulatory Commission Order No. 888, "Promoting Wholesale Competition Through Open-Access Non-Discriminatory Transmission Services by Public Utilities; Recovery of Stranded Costs by Public Utilities and Transmitting Utilities," 75 FERC 61,080, 24 Apr 96. The provisions of Order 888 are explained in more pedestrian language in W. M. Warwick, "A Primer on Electric Utilities, Deregulation, and Restructuring of U.S. Electricity Markets," Federal Energy Management Program, PNNL-13906, May 2002. Both New Hampshire and Rhode Island enacted restructuring legislation earlier but implemented it later. Pennsylvania restructured shortly after California with no negative consequences. See U.S. Energy Information Administration, "Status of State Electric Industry Restructuring Activity," February 2003 (www.eia.doe.gov/cneaf/electricity).

76. See, for example, J. Yardley, "Texas Learns in California How Not to Deregulate," New York Times, 10 Jan 01.

77. B. McLean and P. Elkind, *The Smartest Guys in the Room: The Amazing Rise and Scandalous Fall of Enron* (Portfolio Hardcover, 2003); K. Eichenwald and D. B. Henriques, "Enron Buffed Image to a Shine Even as It Rotted from Within," New York Times, 10 Feb 02. It was never clear exactly how much the California energy crisis contributed to Enron's fall, even though the two events later became synonymous in many people's minds. See A. Berenson, "California May Have Had Big Role in Enron's Fall," New York Times, 9 May 02.

78. H. Kurtz, "The Enron Story That Waited to Be Told," Washington Post, 18 Jan 02.

79. B. McLean, "Is Enron Overpriced?," Fortune, 5 Mar 01.

80. "Enron Net Rose 40% in Quarter," New York Times, 13 Jul 01.

81. R. A. Oppel Jr. and A. Berenson, "Enron's Chief Executive Quits After Only 6 Months in Job," New York Times, 15 Aug 01; P. Krugman, "Reckonings; Enron Goes Overboard," New York Times, 17 Aug 01.

82. K. N. Gilpin, "Enron Reports $1 Billion in Charges and a Loss," New York Times, 17 Oct 01.

83. A. Berenson, "S.E.C. Opens Investigation into Enron," New York Times, 1 Nov 01.

84. R. A. Oppel Jr. and A. R. Sorkin, "Enron Files Largest U.S. Claim for Bankruptcy," New York Times, 3 Dec 01.

85. The full extent of Enron's culpability in California's crisis was never clear. In 2004 lawyers from the Snohomish County Public Utility District, then engaged in a lawsuit with Enron over termination of a $122 million power-delivery contract, released what they claimed to be confidential Enron accounting records showing that Enron had manipulated the markets nearly every day of the crisis. See P. Behr, "Records Show Enron Manipulation, Washington State Utility Says," Washington Post, 15 Jun 04. See also J. Roberts, "Enron Traders Caught on Tape," CBS News, 1 Jun 04; J. Holguin, "More Enron Tapes, More Gloating," CBS News, 8 Jun 04.

86. On May 23, 2001, the *Wall Street Journal* reported the following quote from California attorney Bill Lockyer: "I would love to personally escort [Enron Corp. Chairman Kenneth] Lay to an 8-by-10 cell that he could share with a tattooed dude who says, 'Hi my name is Spike, Honey.'" See J. Warren, "Lockyer Fires Earthy Attack at Energy Exec.," Los Angeles Times, 23 May 01.

87. C. Palmeri, L. Cohn, and W. Zellner, "California's Power Failure," Business Week, 8 Jan 01; L. Bergman and J. Gerth, "Power Trader Tied to Bush Finds Washington All Ears," New York Times, 25 May 01; D. Lazarus, "Memo Details Cheney-Enron Links," San Francisco Chronicle, 30 Jan 02; C. Berthelsen and S. Winokur, "Enron's Secret Bid to Save Deregulation," San Francisco Chronicle, 26 May 01.

88. P. Behr, "Papers Show That Enron Manipulated Calif. Crisis," Washington Post, 7 May 02; "Enron 'Manipulated Energy Crisis,'" BBC News, 7 May 02; R. A. Oppel Jr., "How Enron Got California to Buy Power It Didn't Need," New York Times, 8 May 02.

89. The $1.52 billion settlement was a Pyrrhic victory for California because Enron, still under bankruptcy protection, was paying unsecured creditors only 20 cents on the dollar. See J. Mouawad, "Settlement Is Reached with Enron," New York Times, 16 Jul 05.

90. "Third Annual Report on Market Issues and Performance, January–December 2001," California Independent System Operator (Cal ISO), January 2002.

91. J. Sterngold, "California's New Problem: Sudden Surplus of Energy," New York Times, 19 Jul 01.

92. K. Eichenwald and M. Richtel, "Enron Trader Pleads Guilty to Conspiracy," New York Times, 18 Oct 02; J. R. Wilke and R. Gavin, "Brazen Trade Marks New Path of Enron Probe," Wall Street Journal, 22 Oct 02.

93. E. Kahn, S. Stoft, and T. Belden, "Impact of Power Purchases from Non-Utilities on the Utility Cost of Capital," Utilities Policy **5**, 3 (1995); J. Eto, S. Stoft, and T. Belden, "The Theory and Practice of Decoupling Utility Revenues from Sales," Utilities Policy **6**, 43 (1997); N. Rivera Brooks and N. Vogel, "For Enron's Belden, Success Bred Power," Los Angeles Times, 18 Oct 02.

94. The subordinates in question were Jeffrey Richter and John Forney. See K. Eichenwald, "Second Enron Energy Trader Pleads Guilty," New York Times, 5 Feb 03; L. Johnston, "Former Enron Trader Pleads Guilty," CBS News, 5 Aug 04.

95. *Final Report on Price Manipulation in Western Markets*, Federal Energy Regulatory Commission, Docket No. PA02-2-000, March 2003. The Federal Energy Regulatory Commission maintains an extensive library of material relevant to the California energy crisis at www.ferc.gov/industries/electric/indus-act/wec.asp.

96. The larger of these were Reliant Resources $0.84 billion (106 FERC 61,207) 4 Mar 04; Morgan Stanley $0.86 billion (106 FERC 61,237) 8 Mar 04; Powerex $1.30 billion (106 FERC 61,304) 26 Mar 04;Williams $0.14 billion (108 FERC 61,002) 2 Jul 04; Dynegy $0.28 billion (109 FERC 61,071) 25 Oct 04; Duke $0.21 billion (190 FERC 61,257) 7 Dec 04; Enron $1.48 billion (113 FERC 61,171) 15 Nov 05; Reliant $0.51 billion (113 FERC 61,308) 22 Dec 05. A $0.43 billion settlement with Mirant 13 Apr 05 became moot when Mirant went bankrupt. See C. Berthelsen and D. R. Baker, "Mirant Claims Settled / Bankrupt Energy Company to Forgo $320 Million," San Francisco Chronicle, 15 Jan 05. The Enron settlement became moot for the same reason. How much of the money has actually been paid remains unclear. See letter from U.S. senator Dianne Feinstein to FERC Chairman Jon Wellinghoff, dated 11 May 09 (see note 95).

97. The pertinent wholesale price figures as well as the specific terms "economic withholding," "noncompetitive bidding," and "unpersuasive" appear on pp. VI-45 through VI-52 of FERC PA02-2-000 (see note 95). The companies in question were Bonneville Power Administration, Dynegy, Enron, Idaho Power, Los Angeles Department of Water and Power, Miriant, Powerex, Reliant, and Williams.

98. The California ISO reported an average wholesale price of $330 per megawatt-hour during December 2000. See "Second Annual Report on Market Issues and Performance," California Independent System Operator, November 2001. The spot price was higher than this. For example, the British Columbia Power Exchange Corporation (Powerex) bid an average price of $750 per megawatt-hour on 12 Dec 00, with three bids in excess of $1,100. See FERC PA00-2-000, p. VI-50 (see note 95).

99. S. H. Verhovek, "Energy Secretary Rejects Request to Cap Electricity Rates," New York Times, 3 Feb 01.

100. The only well-documented case of physical withholding occurred in June 2000, when Reliant Energy ordered the shutdown of one of its generators during a power emergency. See E. Douglass, "U.S. Seeks to Indict Reliant Resources Unit," Los Angeles Times, 9 Mar 04. Evidence for a second case in January 2001 was

found on telephone tapes discovered serendipitously in an Enron warehouse. See T. Egan, "Tapes Show Enron Arranged Plant Shutdown," New York Times, 4 Feb 05.

101. P. Davidson, "Shocking Electricity Prices Follow Deregulation," USA Today, 10 Aug 07.

102. "Energy in Developing Countries," U.S. Congress, Office of Technology Assessment, OTA-E-486, January 1991.

103. Below are gross domestic product and energy consumption figures for the thirty largest economies. The first figure is the 2009 GDP as reported in the 2011 CIA World Factbook, expressed as a multiple of 10^{12} 2009 U.S. dollars. The second figure is the 2009 primary energy use reported in the 2010 BP Statistical Review of World Energy, expressed as a multiple of 10^8 TOE (tonnes of oil equivalent): United States 14.25, 21.87; China 9.14, 21.77; Japan 4.15, 4.64; India 3.68, 4.69; Germany 2.84, 2.90; Russia 2.14, 6.35; Brazil 2.02, 2.26; United Kingdom 2.15, 1.99; France 2.11, 2.42; Italy 1.75, 1.63; Mexico 0.89, 1.63; South Korea 1.38, 2.38; Spain 1.37, 1.33; Canada 1.29, 3.19; Indonesia 0.97, 1.28; Turkey 0.89, 0.93; Australia 0.86, 1.19; Iran 0.81, 2.05; Taiwan 0.74, 1.06; Poland 0.69, 0.92; Netherlands 0.66, 0.93; Saudi Arabia 0.60, 1.92; Argentina 0.55, 0.73; Thailand 0.54, 0.95; South Africa 0.51, 1.27; Egypt 0.47, 0.76; Pakistan 0.44, 0.66; Colombia 0.42, 0.29; Malaysia 0.39, 0.56. The U.S. numbers give a GDP per joule of $\$14.25 \times 10^{12} / (21.82 \times 10^8 \text{ TOE} \times 4.20 \times 10^{10} \text{ joules/TOE} = \$1.55 \times 10^{-7} \text{ joule}^{-1}$. The 2009 retail cost of electricity in the United States was roughly $\$0.10 \text{ kwh}^{-1} / (10^3 \text{ watts/kilowatt} \times 3,600 \text{ sec/hour}) = \$2.78 \times 10^{-8} \text{ joule}^{-1}$ (in 2009 U.S. dollars). The U.S. GDP per joule is 5.58 time this latter number. The calculation for other countries is similar.

CHAPTER 4: CARBON FOREVER

104. There is strong evidence in the geologic record that biological activity created earth's oxygen atmosphere. See A. D. Anbar *et al.*, "A Whiff of Oxygen Before the Great Oxidation Event?" Science **317**, 1903 (2007); A. J. Kaufman *et al.*, "Late Archaen Biosphere Oxygenation and Atmospheric Evolution," Science **317**, 1900 (2007); R. Buick, "When Did Oxygenic Photosynthesis Evolve?," Phil. Trans. R. Soc. B **363**, 2731 (2008).

105. They don't disappear immediately, for incomplete burning results in carbon-based combustion products that remain in the atmosphere as pollutants until natural processes such as photolysis or biodegradation complete their conversion to carbon dioxide. See D. Mackay *et al.*, *Handbook of Physical-Chemical Properties and Environmental Fate for Organic Chemicals* (CRC Press, 2006); O. Meyer and H. G. Schlegel, "Biology of Aerobic Carbon Monoxide-Oxidizing Bacteria," Annu. Rev. Microbiol. **37**, 277 (1983).

106. The classic work on this subject is J. C. Slater, *Quantum Theory of Molecules and Solids, Vols. I–IV* (McGraw-Hill, 1967). See also F. Jensen, *Introduction to*

Computational Chemistry (Wiley, 2007); R. M. Martin, *Electronic Structure: Basic Theory and Practical Methods* (Cambridge U. Press, 2004).

107. The quantum unit of length, the bohr radius a_0, is given in terms of the electron charge e, the electron mass m, the electric permittivity of free space ϵ_0, and planck's constant $\hbar$ by $a_0 = 4\pi\epsilon_0\hbar^2/me^2 = 4\pi \times 8.84 \times 10^{-12}$ farads/m $\times$ $(1.055 \times 10^{-34}$ joule-sec $/ 1.602 \times 10^{-19}$ coulombs$)^2 / 9.11 \times 10^{-31}$ kg $= 5.29 \times 10^{-11}$ meters. Here are some relevant bond lengths: H-H 0.74 1.40, C-C 1.54 2.91, H-O 0.96 1.81, C-H 1.09 2.06, C-O 1.43 2.70, O-O 1.48 2.80, O=O 1.21 2.29, C=O 1.20 2.27. The first number is the length expressed as a multiple of 10^{-10} meters. The second is the length in bohr radii. See L. Pauling, *The Nature of the Chemical Bond* (Cornell U. Press, 1960); J. A. Bell, *Chemistry* (W. H. Freeman, 2004).

108. J. J. Katz *et al.*, *The Chemistry of the Actinide and Transactinide Elements*, Vols. I–V (Springer, 2007).

109. The quantum unit of energy, the rydberg, is the companion of the bohr (see note 107): $R_\infty = e^2 / (8\pi\epsilon_0 a_0) = (1.602 \times 10^{-19}$ coulombs$)^2 /(8\pi \times 8.85 \times 10^{-12}$ farads/m $\times 5.29 \times 10^{-11}$ m$) = 2.18 \times 10^{-18}$ joules. The corresponding macroscopic energy is 2.18×10^{-18} joules $\times 6.022 \times 10^{23}$ mole$^{-1} = 1.31 \times 10^6$ joules/mole. Here are some relevant bond energies (enthalpies): H-H 4.36 0.332, C-C 3.47 0.264, H-O 4.59 0.350, C-H 4.13 0.315, C-O 3.60 0.274, O-O 1.42 0.108, O=O 4.98 0.379, C=O 7.98 0.608. The first number is the macroscopic energy expressed as a multiple of 10^5 joules/mole. The second is the energy in rydbergs per bond. Note that these numbers vary at the % level from one reference to another. The energy released in burning is given roughly by appropriate sums and differences of them, per Hess's law. For example, for solid carbon burning to carbon dioxide (C + $O_2 \rightarrow CO_2$) we have, in multiples of 10^5 joules per mole of carbon, $2 \times 7.98 - 2 \times 3.47 - 4.98 = 4.04$. The measured number is 3.94×10^5 joules/mole. Here are some relevant measured heats of formation (enthalpies to make substances from their elements): H_2 0.0 0.0, O_2 0.0 0.0, C (graphite) 0.0 0.0, CO_2 −3.94 −0.300, H_2O (liquid) −2.85 −0.217, CH_4 −0.75 − 0.057, C_7H_{16} −1.88 −0.144. The first number is the energy expressed as a multiple of 10^5 joules per mole. The second is the energy in rydbergs per molecule. See NIST Chemistry WebBook.

110. Increasing the pressure on a material by δp decreases its volume V by an amount δV satisfying $\delta V/V = -\delta p/K$. The material's bulk modulus K has units of pressure (see notes 107, 109). The quantum unit of pressure is $R_\infty/a_0^3 = 2.18 \times 10^{-18}$ joules $/ (5.29 \times 10^{-11}$ m$)^3 = 1.47 \times 10^{13}$ pascals (1.46×10^8 atmospheres). The shear modulus G, which describes the resistance of the material to twisting, is comparable to K. Here are some measured values of K and G, expressed as a multiple of 10^{10} pascals: C (diamond) 44.2 47.8, Fe 17.3 8.14, Cu 14.0 4.75, Al 7.78 2.52, SiO_2 7.85 3.12, NaCl 2.45 1.49, H_2O (ice) 0.935 0.341, C_nH_{2n+2} (solid) 0.340 0.026. Note that these numbers depend on sample morphology and are thus uncertain to 10%. See D. E. Gray, ed., *American Institute of Physics Handbook* (McGraw-Hill, 1957), pp. 2–56 and 3–80; H. J. McSkimin and P. Andreath, "Elastic Moduli of Diamond as a Function of Pressure and Temperature," J. Appl. Phys. **43**, 2944 (1972); P. H.

Gammon, H. Kiefte, and M. J. Clouter, "Elastic Constants of Ice Samples by Brillouin Spectroscopy," J. Phys. Chem. **87**, 4025 (1983).

111. Gasoline's behavior at 10^{11} pascals (1,000,000 atmospheres) is not completely understood, even though such pressures can be generated in the laboratory, because of the tendency of alkanes to phase separate at these pressures into hydrogen and diamonds. See M. Ross, "The Ice Layer in Uranus and Neptune—Diamonds in the Sky?" Nature **292**, 435 (1981); A. Zerr *et al.*, "Decomposition of Alkanes at High Pressure and Temperature," High Pressure Research **26**, 23 (2006). The representative bulk modulus of gasoline in the range of interest is about K = 6×10^{10} pascals. Assuming that gasoline's specific gravity is 0.719, compressing gasoline to twice its normal density stores energy ($\ln 2 - 1/2$) K/ρ = 0.193 × 6.0 × 10^{10} joules/m^3 / 719 kg/m^3 = 1.61 × 10^7 joules/kg. The combustive energy of gasoline is 4.4 × 10^7 joules/kg.

112. The maximum detonation pressure of chemical high explosive is about 5 × 10^{10} pascals (500,000 atmospheres). See R. Meyer, J. Köhler, and A. Homburg, *Explosives* (Wiley-VCH, 2007).

113. A ten-gallon tank of gasoline compressed to half its volume would contain compressive energy (see note 111) 10 gals × 791 kg/m^3 × 1.61 × 10^7 joules/kg / 264.2 gals/m^3 = 4.8 × 10^8 joules. The energy in a stick of dynamite is about 2.1 × 10^6 joules. See H. P. Gillette, *Rock Excavation* (M. C. Clark, 1904).

114. The environment is very good at getting rid of carbon-based combustion products, many of which are poisonous. A good example is ordinary soot, which contains a mix of molecules called polycyclic aromatic hydrocarbons that are structurally similar to benzene, a notorious carcinogen. See J. C. Fetzer, *The Chemistry and Analysis of the Large Polycyclic Aromatic Hydrocarbons* (Wiley, 2000).

115. The explosive range of H_2 in air is 0.04 to 0.75 volume fraction. The explosive range of CH_4 is 0.05 to 0.15 volume fraction. See D. P. Nolan, *Handbook of Fire and Explosion Protection Engineering Principles for Oil, Gas, Chemical and Related Facilities* (William Andrew, 1997). The current market price of hydrogen is difficult to pin down. Various anecdotal sources on the Internet place it at about $100/kg. In 2001 NASA quoted the price of liquid hydrogen used in its rockets at 98 cents per gallon or $0.98/gal × 264.2 gal/m^3 / 79 kg/m^3 = $3.28/kg. See "Space Shuttle Use of Propellants and Fluids," National Aeronautics and Space Administration, FS-2001-09-015-KSC, September 2001. A recent study by the National Academy of Engineering projected that hydrogen could eventually be delivered as transportation fuel for $2.11/kg. However, this low price would be achieved only if the hydrogen were made from fossil fuel. See *The Hydrogen Economy: Opportunities, Costs, Barriers and R&D Needs* (National Academies Press, 2004). At this latter very low price, hydrogen's cost per joule would be comparable to gasoline's.

116. According to the DOE Office of Energy Efficiency and Renewable Energy (www.eere.energy.gov), about 95% of the hydrogen produced today in the United States is made via steam-methane reforming. See also B. Balasubramanian *et al.*, "Hydrogen from Methane in a Single-Step Process," Chem. Eng. Sci. **54**, 3543 (1999).

117. The combustive energy of hydrogen gas is 0.217 ry/molecule $\times$ 2.18 $\times$ 10^{-18} joules/ry $\times$ 6.022 $\times$ 10^{23} molecules/mole = 2.86 $\times$ 10^5 joules/mole (see note 109). A tank of hydrogen gas compressed to 2.17 $\times$ 10^7 pascals (215 atmospheres), the typical pressure of a scuba tank, has a combustive energy content, compared to the same volume of gasoline, of (2.17 $\times$ 10^7 joules/m^3 $\times$ 2.86 $\times$ 10^5 joules/mole) / (8.314 joules $mole^{-1}$ $°K^{-1}$ $\times$ 300°K $\times$ 4.40 $\times$ 10^7 joules/kg $\times$ 719 kg/m^3) = 0.079. The same tank filled with cryogenic liquid hydrogen (specific gravity 0.079) has energy content compared to gasoline of (2.86 $\times$ 10^5 joules/mole $\times$ 79 kg/m^3) / (0.002 kg/mole $\times$ 4.40 $\times$ 10^7 joules/kg $\times$ 719 kg/m^3) = 0.36. Hydrogen liquefies at 20°K (−256°C), a temperature cold enough to cause severe burns. Hydrogen can be stored at about twice its liquid density as metal hydride. This has technical difficulties, however, including poisoning of the host metal by impurities and the requirement for heating to dislodge the hydrogen. The storage medium is also heavy because it contains transition metals. The mass density per hydrogen for Mg_2NiH_4, for example, is 4.5 times that of gasoline. See, e.g., L. Schlapbach and A. Züttel, "Hydrogen-Storage Materials for Mobile Applications," Nature **414**, 353 (2001).

118. The amount of hydrogen stored in a tank of gasoline is actually greater than the amount stored in a tank of liquid hydrogen. The ratio of the two is 16/(12 $\times$ 7 + 16) $\times$ 719 kg/m^3 / 79 kg/m^3 = 1.46.

119. For all of these technologies, the maximum storage density per unit mass is roughly σ_y/ρ, where σ_y is the relevant material's yield strength and ρ is its mass density. In each case we reach this conclusion by solving $\mathcal{F}_\beta - \sum_{\alpha=1}^{3} \partial\sigma_{\alpha\beta}/\partial x_\alpha = \rho \, \partial^2\psi_\beta/\partial t^2$. Here ψ_β represents the local atomic displacement, $\mathcal{F}_\beta$ represents any externally applied force, and $\sigma_{\alpha\beta} = G(\partial\psi_\beta/\partial x_\alpha + \partial\psi_\alpha/\partial x_\beta) + (K - 2G/3)\delta_{\alpha\beta} \sum_{\gamma=1}^{3} \partial\psi_\gamma/\partial x_\gamma$ is the Cauchy stress tensor. K and G are the material's bulk modulus and shear modulus (see note 110). For example, for a spherical tank of inner radius r_1 and outer radius r_2, containing gas at pressure p, we have in the tank wall $\psi_\beta = (p/6K) \, r_1^3/(r_1^3 - r_2^3)\partial/\partial x_\beta \, (r^2 - (3K/2G) \, r_2^3/r)$, and thus $\sigma_{rr} = p \, (1 - r_2^3/r^3) / (1 - r_2^3/r_1^3$) and $\sigma_{\theta\theta} = \sigma_{\phi\phi} = p \, (1 + 0.5 \, r_2^3/r^3) / (1 - r_2^3/r_1^3)$. The transverse components of stress are negative (in tension) and maximized on the inner wall. The condition that the tank not explode is thus $\sigma_y > p \, (r_1^3 + 0.5 \, r_2^3) / (r_2^3 - r_1^3)$. If we take the tank's energy content to be the isothermal compressive energy of the gas it contains, then the energy stored per unit (wall) mass satisfies $E/M < (\sigma_y/\rho) / (1 + 0.5 \, r_2^3/r_1^3)$. The calculation for a spherical flywheel proceeds similarly except that there is now a maximum allowed spin rate instead of maximum allowed pressure. The energy per unit mass of the flywheel satisfies $E/M < 7/8 \times (3K + 4G)/3G \times \sigma_y/\rho$. For a twisted rod (i.e., a spring), we find that the rod can't store twist energy without breaking unless it is big. The energy per unit mass of the rod satisfies $E/M < \sigma_y^2/4G\rho$.

120. A material's maximum yield strength σ_y is about 100 times smaller than its shear modulus G (see note 119). It isn't possible to be more precise than this, because strength is highly sensitive to chemical impurities and deformation history. Some measured values of σ_y and σ_y/ρ expressed as multiples of 10^9 pascals and 10^5

joules/m^3, respectively: C (Fiber Composite) 1.4 8.8, SiO$_2$ (Fiberglass) 1.1 5.8, Ti (Ti-6Al-4V) 0.83 1.8, Fe (A514) 0.62 0.79, Al (7075) 0.50 1.9, Cu (C81300) 0.25 0.28, Polyethylene 0.036 0.039. See M. J. Donachie Jr., *Titanium: A Technical Guide*, 2nd ed. (ASM, 2000); C. G. Salmon, J. E. Johnson, and F. A. Malhas, *Steel Structures: Design and Behavior* (Prentice Hall, 2008); J. R. Davis, *Aluminum and Aluminum Alloys* (ASM, 1993); J. R. Davis, *Copper and Copper Alloys* (ASM, 2001); L. H. Sperling, *Introduction to Physical Polymer Science* (Wiley, 2005); J. F. Shackelford and W. Alexander, eds., *CRC Materials Science and Engineering Handbook*, 3rd ed. (CRC Press, 2001); S. M. Lee, ed., *Handbook of Composite Reinforcements* (Wiley-VCH, 1992); J. M. F. de Pavia, S. Mayer, and M. C. Rezende, "Comparison of Tensile Strength of Different Carbon Fabric Reinforced Epoxy Composites," Mat. Res. **9**, 83 (2006); J. M. Corum *et al.*, "Basic Properties of Reference Crossply Carbon-Fiber Composite," Oak Ridge National Laboratory, ORNL/TM-2000/29, February 2000.

121. Yield strengths of practical materials are lower than the ideal values for perfect crystals because imperfections focus stress, thus enabling the material to fail by moving one atom at a time rather than all the atoms at once. In metallurgy this effect is called dislocation motion, and it's implicated in plastic flow, embrittlement, and fracture. See W. F. Hosford, *Mechanical Behavior of Materials* (Cambridge U. Press, 2005). In the case of metals, reducing the sample size to very small dimensions increases the yield strength enormously by enabling dislocations to find their way out of the sample quantum mechanically and disappear. See C. Herring and J. K. Galt, "Elastic and Plastic Properties of Very Small Metal Specimens," Phys. Rev. **85**, 1060 (1952); J. Franks, "Metal Whiskers," Nature **177**, 984 (1956). This strengthening effect of smallness has led to the hope that nanoscale composite materials might reach yield strengths much greater than those of steel. See E. W. Wong, P. E. Sheehan and C. M. Lieber, "Nanobeam Mechanics: Elasticity, Strength and Toughness of Nanorods and Nanotubes," Science **277**, 1971 (1997); P. J. F. Harris, "Carbon Nanotubes and Related Structures: New Materials for the Twenty-First Century," Am. J. Phys. **72**, 415 (2004); S. Stankovich *et al.*, "Graphene-Based Composite Materials," Nature **442**, 282 (2006).

122. Diamond anvil cells now routinely achieve pressures of 6×10^{10} pascals (0.6 million atmospheres) and can be coaxed into reaching pressures several times greater than this with some patience and an ample budget for replacing broken diamonds. See D. J. Weidner, Y. Wang, and M. T. Vaughn, "Strength of Diamond," Science **266**, 419 (1994); A. F. Goncharov *et al.*, "Compression of Ice to 210 Gigapascals: Infrared Evidence for a Symmetric Hydrogen-Bonded Phase," Science **273**, 218 (1996); F. Occelli, P. Loubeyre, and R. LeToullec, "Properties of Diamond Under Hydrostatic Pressure up to 140 GPa," Nature Mat. **2**, 151 (2003); Y. Akahama and H. Kawamura, "Pressure Calibration of Diamond Anvil Raman Gauge to 310 GPa," J. Appl. Phys. **100**, 043516 (2006).

123. A typical scuba tank stores air at 2.15×10^7 pascals (215 atmospheres) in a volume of 0.011 m^3. On the rare occasions that one explodes, it thus releases energy

2.17×10^7 joules/m^3 × 0.011 m^3 × ln(215) = 1.28×10^6 joules, or about 1/2 stick of dynamite (see note 113). The pressure carried by a paintball canister is 2.57×10^7 pascals (257 atmospheres). The pressure in a modern supercritical steam boiler is 2.45×10^7 pascals (245 atmospheres). The pressure at which ammonia is synthesized in the industrial Haber process is 2.0×10^7 pascals (200 atmospheres). Even though they were pressurized to less than 10^6 pascals (10 atmospheres), nineteenth-century steam boilers did tremendous damage when they failed—in some cases destroying entire buildings. See R. H. Thurston, *Steam Boiler Explosions in Theory and Practice* (Wiley, 1903). The failure of large flywheels similarly demolished suites of rooms and blew holes in brick walls. See W. H. Boehm, *Fly-Wheel Explosions* (Fidelity Casualty Co. of New York, 1915), available at www.rustyiron .com. Lurid videos of flywheel explosions in race cars are posted at various places on the Internet.

124. C. Knowlen *et al.*, "High Efficiency Energy Conversion Systems for Liquid Nitrogen Automobiles," SAE Trans. **107**, 1837 (1998). Note that the energy densities in this paper don't include the weight of the tank (see note 125).

125. A cryogenic liquid yields the maximum mechanical energy if it is heated to room temperature (T = 300°K) at constant volume and then expanded isothermally in an engine. This is measured quantity implicit in published equations of state. Here are some typical values, expressed as a multiple of 10^5 joules/kg: N_2 6.69, O_2 4.90, F_2 4.46, CH_4 8.65, NH_3 4.87, C_2H_6 3.41, CH_3F 3.30, H_2S 2.95. See E. W. Lemmon, M. G. McLinden, and P. G. Freund, "Thermophysical Properties of Fluid Systems," in *NIST Chemistry WebBook, NIST Standard Database Number 69*, P. J. Linstrom and W. G. Mallard, eds. (National Institute of Standards and Technology, 2008), http://webbook.nist.gov. The relevant isotherm passes through the supercritical or solid regions of the phase diagram and so cannot be inferred reliably from low-pressure measurements. Nonetheless, the measured deliverable energy W is close to the Clausius-Clapeyron value $W \simeq \Delta H_{vap} (T/T_{boil} - 1)$ where T_{boil} and ΔH_{vap} are the boiling temperature and enthalpy of vaporization at one atmosphere.

126. A gas compressed to liquid at room temperature (T = 300°K) yields maximum mechanical energy when gas is drawn off isothermally at the liquid's vapor pressure and expanded isothermally in an engine. Here are some typical values, expressed as a multiple of 10^5 joules/kg: CO_2 2.14, NH_3 3.49, C_2H_6 3.70, CH_3F 2.77, H_2S 2.25. The measured energy is close to the ideal gas value $W \simeq (RT/M_{mol}) \ln (p_{vap}/p_0)$, where M_{mol} is molar mass, p_{vap} is the vapor pressure, and p_0 is one atmosphere.

127. The storage capacities of batteries cover a range, much the way reported material strengths do, and thus are difficult to pin down precisely. Here are some typical values, expressed as a multiple of 10^5 joules/kg: Pb-Acid 1.26, Ni-Cd 1.44, NiMH 2.52, Li-ion 4.32, Alkaline 5.94, Na-S 6.12. The conversion factor from watt-hours/kg to joules/kg is 1 wh = 3.60×10^3 joules. See R. H. Rashid, *Power Electronics*

Handbook (Academic, 2006). The capacity of alkaline batteries is from the Duracell website.

128. Lithium atoms are lighter than either carbon or oxygen atoms, and this lightness also gives them unique quantum mechanical properties important for battery design. See G. A. Nazri and G. Pistoia, *Lithium Batteries: Science and Technology* (Springer, 2009).

129. The storage capacity of a high-end lithium-ion battery is 7.2×10^5 joules/kg. The storage capacity of gasoline is 4.4×10^7 joules/kg. A conventional automobile engine is 25% efficient, so the energy content of gasoline is effectively only 1.1×10^7 joules/kg. Electric cars also have inefficiencies, but they aren't as well documented. See M. H. Westbrook, *The Electric Car: Development and Future of Battery, Hybrid and Fuel-Cell Cars* (Institution of Engineering and Technology, 2001); D. Linden and T. B. Reddy, eds., *Handbook of Batteries* (McGraw-Hill, 2001). Batteries also have a serious cost problem. A lithium-ion battery, for example, costs about U.S. \$150 per kwh to manufacture and can undergo about 1000 charge-discharge cycles before it must be replaced. The total storage cost per joule is thus \$150/kwh $/ (3.6 \times 10^6$ joules/kwh) $= \$4.17 \times 10^{-5}$ joule^{-1}. The pump price of gasoline is presently \$3.00/gal $/ (2.81$ kg/gal $\times 4.4 \times 10^7$ joules/kg) $= \$2.02 \times 10^{-8}$ joule^{-1}.

130. E. P. Murray, T. Tsai, and S. A. Barnett, "A Direct Methane Fuel Cell with a Ceria-Based Anode," Nature **400**, 649 (1999).

131. R. O'Hayre *et al.*, *Fuel Cell Fundamentals* (Wiley, 2009).

132. Much present development in methanol fuel cells is unfortunately secret for either military or industrial competitiveness reasons. See M. Williams, "Sharp Reveals Progress in Consumer-Use Fuel Cells," PC World, 15 May 08; K. Bourzac, "More-Powerful Fuel Cells," Technology Review, 22 May 08; R. C. Johnson "Direct-Methanol Fuel Cell Takes DoD Prize," EETimes, 6 Oct 08; "Ultracell to Build Fuel Cell Systems for Unmanned Air Vehicles," MSNBC, 29 Jun 09. There are also serious efforts to make fuel cells burn ethanol. See K. Bullis, "Efficient Ethanol Fuel Cells," Technology Review, 2 Feb 09.

133. C. Barras, "Breathing Batteries Could Store 10 Times the Energy," New Scientist, 19 May 09; K. Bourzac, "IBM Invests in Battery Research," Technology Review, 11 Jun 09; M. Fischer, M. Werber, and P. V. Schwartz, "Batteries: Higher Energy Density Than Gasoline?" Energy Policy **37**, 2639 (2009).

134. A fuel cell locomotive was recently built for Burlington Northern-Santa Fe, thanks to the support of Kansas senator Sam Brownback. See A. M. Bush, "New Locomotive Unveiled," Topeka Capital-Journal, 29 Jun 09. Fuel cell forklifts were recently deployed at the Anheuser-Busch facility in Ft. Collins, Colorado. See S. Porter, "Fuel Cells to Power A-B Forklifts," Northern Colorado Business Report, 5 Jun 09. See also J. E. Spiegel, "For Hartford, a Fuel-Cell Bus Milestone," New York Times, 15 Apr 07; M. L. Wald, "US Drops Research into Fuel Cells for Cars," New York Times, 7 May 09; P. Olson, "Collision Course," Forbes, 29 Jun 09.

135. The manufacturing cost of an engine is a tightly held secret of the automotive industry and thus difficult to document. A value of $600 for a 4-cylinder engine is mentioned in V. Reitman, "Toyota Introduces New Engine in U.S. That Is Much Lighter, Cheaper to Make," Wall Street Journal, 30 Oct 97. This would amount to 5% of a $13,000 compact car sales price. Compact sales prices were closer to $17,000 in 1997, so the 5% estimate may be high. Rebuilt automobile engines presently retail between $500 and $2,000.

136. Assuming the "horse" delivering one horsepower to be a Clydesdale, the mass of which is about 1000 kg, the power per unit mass of a horse comes out to be 1 hp × 746 watts/hp / 1,000 kg = 0.746 watts/kg. Automobile engines typically provide 400 watts/kg and can be made to provide more with such tricks as supercharging.

137. The takeoff speed of a Boeing 747 is 81 m/sec. Each of its four GE CF6-80C2B1 engines delivers 2.6×10^5 newtons of thrust, maximum throttle, and has a mass of 4.3×10^3 kg (www.geae.com). The power delivered per kilogram of engine mass on liftoff is thus 2.6×10^5 newtons × 81 m/sec / 4.3×10^3 kg = 4.90×10^3 watts/kg. The power per unit engine mass is actually larger than this but difficult to pin down because it depends on the speed and altitude of the plane. The fuel consumption, however, is straightforward. Various anecdotal reports on the Internet say that a 747 on its takeoff roll burns fuel at about 10 kg/sec. The power consumed per kilogram of engine mass during the roll is thus 10 kg/sec × 4.4×10^7 joules/kg / ($4 \times 4.3 \times 10^3$ kg) = 2.56×10^4 watts/kg. The official thrust and specific impulse ratings of the engine give the slightly higher value of 1.71×10^{-5} kg/newton-sec × 2.6×10^5 newtons × 4.4×10^7 joules/kg / 4.3×10^3 kg = 4.55×10^4 watts/kg. See also www.geae.com.

138. See the discussion of peaking plants in F. Lévêque, *Competitive Electricity Markets and Sustainability* (Edward Elgar, 2007).

139. The Fischer-Tropsch process was invented in Germany in 1925. It was subsequently perfected by the Nazi government and used as one of two technologies for achieving "petroleum independence" during World War II. Fischer-Tropsch production peaked in 1944 at 4.1 million bbl per year. Fischer-Tropsch technology was then adopted by the South African government and further refined in the late 1970s as a means of evading the "voluntary" oil embargo imposed upon it by the United Nations General Assembly. See A. Stranges, "Germany's Synthetic Fuel Industry, 1927–1945," in *The German Chemical Industry in the Twentieth Century*, J. E. Lesch, ed. (Springer, 2000); M. E. Dry, "High-Quality Diesel Via the Fischer-Tropsch Process—A Review," J. Chem. Tech. Biotech. **77**, 43 (2002); B. H. Davis and M. L. Occelli, eds., *Fischer-Tropsch Synthesis, Catalysts, and Catalysis* (Elsevier, 2006). Prof. A. Stranges of Texas A&M University maintains a comprehensive Fischer-Tropsch archive at www.fischer-tropsch.org.

140. How the world economy deals with energy price increases was examined particularly carefully after the oil shocks of the 1970s. The overall picture is that price increases depress gross domestic product significantly and that the easy-

money policies required to counteract this effect are inflationary. See S. P. A. Brown and M. K. Yücel, "Energy Prices and Aggregate Economic Activity: An Interpretive Study," Quarterly Revue of Economics and Finance **42**, 193 (2002); R. E. Hall and K. A. Mork, "Energy Prices, Inflation and Recession 1974–75," Energy J. **1**, 31 (1980).

141. It's also confused in the economics literature by virtue of being politically sensitive and difficult to document. See M. Feldstein, "America Will Fall Harder If Oil Prices Rise," Financial Times, 3 Feb 06.

142. According to the U.S. Energy Information Administration, the fraction of U.S. gross domestic product dedicated to energy expenditures has varied between 6 and 15% over the past four decades. See "Annual Energy Outlook 2009," U.S. Energy Information Administration, DOE/EIA-0383, March 2009.

143. The origin of this problem is that the gasification reaction is endothermic and requires burning of additional carbon to provide the necessary energy. The full synthesis reaction may be idealized as $2\,C + 2\,H_2O + 1/2\,O_2 \rightarrow CO + 2\,H_2 + CO_2 \rightarrow (CH_2) + H_2O + CO_2$. Thus, the amount of carbon dioxide generated for a given amount of transport energy is roughly twice that produced by burning petroleum.

144. Royal Dutch Shell has commercial gas-to-liquids (using natural gas as feedstock) Fischer-Tropsch plants in Bintulu, Malaysia, and Qatar. Other companies pursuing gas-to-liquids ventures include ExxonMobil, Chevron, Statoil, and ConocoPhillips. See S. Romero, "A New Way to Make Old Diesel," New York Times, 18 Jan 06; S. Reed and A. Aston, "What the U.S. Can Learn from Sasol," Business Week, 27 Feb 06; M. L. Wald, "Search for New Oil Sources Leads to Processed Coal," New York Times, 5 Jul 06.

145. Sasol operates the world's only commercial coal-to-liquids plant in Secuna, South Africa. The exact size of the subsidies received by Sasol is difficult to determine, for it was a tightly guarded state secret during apartheid, and Sasol's finances are still very opaque. See A. M. Rosenthal, "The Secret Pipeline," New York Times, 19 Apr 90. A lower bound on the true cost is provided by Canada's oil sands industry, which expanded vigorously in the 1990s, whereas Sasol's coal-to-liquids fuel business did not.

146. An example of the problem this can cause is the language in the 1997 U.S. Energy Independence and Security Act that, according to National Defense Magazine, "prevents the Air Force—or any government agency—from buying synthetic jet fuel unless it is proved to emit less carbon dioxide over the life of the substance than currently used petroleum." See B. Wagner, "Market for Synthetic Aviation Fuels Off to a Shaky Start," National Defense, May 2008; E. L. Andrews, "Lawmakers Push for Big Subsidies for Coal Process," New York Times, 29 May 07.

147. A high-end coal gondola rail car holds 1.2×10^5 kg of coal (www.gatx.com). This amount of coal contains energy 120 tonnes $\times\ 3.17 \times 10^{10}$ joules/tonne $= 3.8 \times 10^{12}$ joules and costs 120 tonnes $\times$ \$50/tonne $=$ \$6,000. The cost of a new hybrid compact car, by contrast, is presently \$22,000 plus extras (www.edmunds.com).

The car's lifetime fuel cost is 10^5 miles × \$3/gal / 55 miles/gal = \$5,400. The car's lifetime energy consumption is 10^5 miles × 2.72 kg/gal × 4.4 × 10^7 joules/kg / 55 miles/gal = 2.18 × 10^{11} joules.

148. The subject of transport fuel from biomass is extremely hot in academic circles at the moment and has a correspondingly large literature. Representative publications: M. J. A. Tijmensen *et al.*, "Exploration of the Possibilities for Production of Fischer Tropsch Liquids and Power Via Biomass Gasification," Biomass and Bioenergy **23**, 129 (2002); G. P. Towler, A. R. Oroskar, and S. E. Smith, "Development of a Sustainable Liquid Fuels Infrastructure Based on Biomass," Environ. Prog. **23**, 334 (2004); J. R. Rostrup-Nielson, "Making Fuels from Biomass," Science **308**, 1421 (2005); P. Fairley, "Growing Biofuels: New Production Methods Could Transform the Niche Industry," Technology Review, 23 Nov 05; H. Ledford, "Liquid Fuel Synthesis: Making It Up as You Go Along," Nature **44**, 677 (2006); A. Faaij, "Modern Biomass Conversion Technologies," Mitigation and Adaptation Strategies for Global Change **11**, 335 (2006); S. R. Schill, "The Fischer-Tropsch/Fat Connection," Biomass Magazine, 13 Nov 07; E. van Steen and M. Claeys, "Fischer-Tropsch Catalysts for the Biomass-to-Liquids (BTL) Process," Chemical Engineering and Technology **31**, 655 (2008); "Rentech Advances Bio-Energy Strategy Through Acquisition of SilvaGas and Investment in ClearFuels," Wall Street Journal, 24 Jun 09.

149. A recent study done at Oak Ridge National Laboratory found that the United States could produce a billion dry tons of biomass annually—about the amount of coal the United States presently consumes every year—without expanding the amount of agricultural land or impinging on the food supply. This could replace about one-third of present-day U.S. fossil fuel energy use. See R. D. Perlock *et al.*, "Biomass as Feedstock for a Bioenergy and Bioproducts Industry: The Technical Feasibility of a Billion-Ton Supply," Oak Ridge National Laboratory ORNL/TM-2005/66, April 2005; J. U. Nef, "An Early Energy Crisis and Its Consequences," Sci. Am., October 1977, p. 141; C. A. Berg, "Process Innovation and Changes in Industrial Energy Use," Science **199**, 608 (1978).

150. This is obvious in the case of Fischer-Tropsch fuels but less so for ethanol made from cellulose by hydrolysis, for this technology is newer and less well understood. However, recent studies have suggested that these cellulosic technologies are similarly inefficient. See T. Hamilton, "Biofuels vs. Biomass Electricity," Technology Review, 8 May 09; J. E. Campbell, D. B. Lobell, and C. B. Field, "Greater Transportation Energy and GHG Offsets from Bioelectricity Than Ethanol," Science **324**, 1055 (2009); N. Moreira, "Wood-Burning Plants Gain Power," Boston Globe, 5 Aug 06.

151. Whether for this reason or "greenwash" (dabbling in alternate energy to blunt criticism over global warming), the world's major oil companies are all investing seriously in biofuels. See D. R. Baker, "Oil Giant Chevron Gives Biofuel a Try," San Francisco Chronicle, 1 Jun 06; C. Krauss, "Tyson Foods and ConocoPhillips to Produce Diesel Fuel from Animal Fat," New York Times, 17 Apr 07; R. Gold, "BP Jumps into Next-Generation Biofuels with Plans to Build Florida Re-

finery," Wall Street Journal, 19 Feb 09; T. Webb, "Shell Dumps Wind, Solar and Hydro Power in Favor of Biofuels," The Guardian, 17 Mar 09; K. A. Dolan, "Shell's Brash Biofuels Partner," Forbes, 22 Apr 09.

152. D. Pimentel and M. H. Pimentel, eds., *Food, Energy, and Society* (CRC Press, 2007); B. Walsh, "Solving the Biofuels vs. Food Problem," Time, 7 Jan 08; A. L. Young, "Finding the Balance Between Food and Biofuels," Env. Sci. Pollu. Res. **16**, 117 (2009); E. Rosenthal, "Rush to Use Crops as Fuel Raises Food Prices and Hunger Fears," New York Times, 6 Apr 11. There are also allegations that diverting farmland to biofuel production worsens the greenhouse effect. See E. Rosenthal, "Biofuels Deemed a Greenhouse Threat," New York Times, 8 Feb 08; T. Searchinger *et al.*, "Use of U.S. Croplands for Biofuels Increases Greenhouse Gases Through Emissions from Land-Use Change," Science **319**, 1238 (2008); J. Fargione *et al.*, "Land Clearing and the Biofuel Carbon Debt," Science **319**, 1235 (2008).

153. This is not a new idea. See K. Bullis, "Biofuels from Saltwater Crops," Technology Review, 5 Feb 10; J. Mandel, "Farming Fuel in Middle Eastern Salt Marshes," Sci. Am. 19 Jan 10; R. Radulovich, "Let's Use Seaweed as Fuel," Cosmos, 10 Jun 08; S. R. Schill, "The Saltwater Soybean," Biodiesel Magazine, 1 Nov 07.

CHAPTER 5: PIPES OF POWER

154. According to the New York Independent System Operator (NYISO) and Consolidated Edison (www.coned.com), New York City's average power consumption is 5.5×10^{10} kwh/year $\times 10^3$ wh/kwh / (24 hours/day $\times$ 365 days/year) = 6.28 $\times 10^9$ watts. The summer and winter maxima are 1.3×10^{10} watts and 7.8×10^9 watts, respectively. The difference is due mainly to summer use of air conditioning, although the longer nights of winter and use of electric heat when it's cold are also factors. A reasonable guess for the amount of energy used to generate light is half the average. Assuming the (very low) conversion efficiency of 100-watt incandescent lightbulbs, the total light emitted is then 3.14×10^9 watts $\times$ 17.5 lumens/watt / (4π lumens/candle) = 4.37×10^9 candles. See "PlaNYC Progress Report 2009," City of New York, 2009; A. Feuer, "City Dims Lights as Heat Strains the Power Grid," New York Times, 2 Aug 06.

155. The electrical resistivity of good insulators is difficult to measure on account of being sensitive to temperature and impurities. Different grades of polyethylene lie in the range 10^{13}–10^{16} ohm-meters (www.boedeker.com). The resistivity of polyethylene is thus larger than that of aluminum by at least 10^{13} ohm m / 2.82×10^{-8} ohm m = 3.55×10^{20}. See D. R. Lide, *CRC Handbook of Chemistry and Physics*, 90th ed. (CRC Press, 2009). This ratio increases as the temperature is lowered and eventually diverges. A metal is, by definition, any substance that conducts electricity at absolute zero temperature.

156. The solution to the free-space Maxwell's equations, equation $\nabla \cdot \mathbf{E} = 0$, $\nabla \cdot \mathbf{B} = 0$, $\nabla \times \mathbf{E} = -\partial \mathbf{E}/\partial t$, $\nabla \times \mathbf{B} = 1/c^2 \, \partial \mathbf{E}/\partial t$, appropriate to a three-phase transmission line

with three perfectly conducting cylindrical wires oscillating at angular frequency ω and separated by distance d, is $\mathbf{E} = V_0$ Re $(\exp(i\xi) (x,y,0)/(x^2 + y^2) + \exp(i\xi + i2\pi/3)$ $(x{-}d,y,0)/((x{-}d)^2 + y^2) + \exp(i\xi - i2\pi/3) (x{+}d,y,0)/((x{+}d)^2 + y^2))$, where $\xi = \omega(z/c -$ t). For the specific case of a 765-kilovolt line, the highest presently used in North America, with d = 16 meters and wires of radius a = 0.015 meters, the prefactor is $V_0 = (2/3)^{1/2} \times 7.65 \times 10^5$ volts / ln(16 m / 0.015 m) = 8.96×10^4 volts. The wire radius is typical for high-voltage applications and is slightly larger than the skin depth of aluminum at 60 Hz (see note 155): c $(2\varepsilon_0\rho/\omega)^{1/2} = 3.0 \times 10^8$ m/sec $\times$ (2 $\times$ 8.85 $\times$ 10^{-12} farads/m $\times 2.82 \times 10^{-8}$ ohm m / $(2\pi \times 60\ \text{sec}^{-1}))^{1/2}$ = 0.011 meters. The total average power carried by the transmission line (all three wires) is then $3\pi c\varepsilon_0 V_0^2$ ln(d/a) = $3\pi \times 3 \times 10^8$ m/sec $\times 8.85 \times 10^{-12}$ farads/m $\times$ (8.96×10^4 volts)^2 $\times$ ln(10 m/0.015 m) = 1.40×10^9 watts. Hydro Quebec quotes a value of 2×10^9 watts. See L. L. Grigsby, ed., *Electric Power Generation, Transmission, and Distribution* (CRC Press, 2007).

157. Engineers usually explain this relationship, which applies to both AC and DC lines, as a stability condition. One wants the power P delivered by the line to increase when the load resistance R_L decreases (i.e., when somebody switches on a light). The equations describing this event are V + δV = (I + δI)(R_L + δR) and δV = $-R_0\ \delta$I. The first equation is just ohm's law with δR representing the load change. The second is the condition that this change should launch a wave-like disturbance backward down the line, a process characterized by the impedance R_0. In the specific case of the 765-kilovolt line above (see note 156), we have R_0 = ln(d/a) / $(\pi\varepsilon_0 c)$ = ln(16 m/0.015 m) / ($\pi \times 8.85 \times 10^{-12}$ farads/m $\times 3 \times 10^8$ m/sec) = 836 ohms. For an undersea cable it is smaller, roughly 30 ohms. For the power, we then have P = (I + δI)(V + δV) and thus δP/P = $(R_0 - R_L)/(R_0 + R_L) \times \delta$R/ R_L. The stability condition is then $R_L > R_0$.

158. The conversion factor from volts to electron pressure depends on the specific metal in question. For the case of aluminum, the maximum electron pressure carried by a 765-kilovolt transmission line is neV = 2.5×10^{28} m$^{-3} \times 1.602 \times 10^{-19}$ coulombs $\times (2/3)^{1/2} \times 7.65 \times 10^5$ volts = 2.5×10^{15} pascals (2.5×10^{10} atmospheres). The electron density *n* is that inferred from the Hall effect. See E. U. Condon and H. Odishaw, *Handbook of Physics* (McGraw-Hill, 1958).

159. The power required by a Boeing 474 on its takeoff roll is 4.4×10^8 watts (see note 137).

160. As of January 1, 1998, the U.S. Environmental Protection Agency limits fuel dispensing rates at filling stations to 10 gallons per minute or less. Anecdotal evidence indicates the actual rate is between 5 and 10 gallons per minute. The Boeing 747 full-throttle consumption rate thus exceeds the compromise filling station pump rate of 7.5 gallons per minute by the factor 10 kg/sec $\times$ 60 sec/min $\times$ 264.2 gal/m^3 / (7.5 gal/min $\times$ 791 kg/m^3) = 26.7.

161. A typical example is the Pacific Connector pipeline, presently planned for a capacity of 10^9 ft^3/day / ((3.28 ft/m)^3 $\times$ 24 h/day $\times$ 3,600 sec/h) = 328 m^3/sec. The

energy equivalent of this flow is 328 m³/sec × 0.67 kg/m³ × 5.5 × 10⁷ joules/kg = 1.21 × 10¹⁰ watts (www.pacificconnectorgp.com). The Ruby pipeline (under construction) and Kern pipeline have capacities 1.5 and 1.9 times this value, respectively. The South Stream pipeline from Russia to central Europe via the Black Sea presently has a planned capacity of 6.8 × 10¹⁰ m³/year / (365 d/y × 24 h/day × 3,600 sec/h) = 2,000 m³/sec, which corresponds to 7.38 × 10¹⁰ watts, or six times that of the Pacific Connector. (Europe currently consumes 6 × 10¹¹ m³ of natural gas per year.) See G. Chazan, "Russia, Italy to Double Capacity of Gas Pipeline," Wall Street Journal, 16 May 09; "El Paso Plans Massive Natural Gas Pipeline," Denver Business Journal, 3 Dec 07.

162. The Alaska Pipeline, for example, has a capacity of 2.0 × 10⁶ bbl/day × 139 kg/bbl × 4.4 × 10⁷ joules/kg / (24 h/day × 3,600 sec/h) = 1.42 × 10¹¹ watts. It delivered this flow reliably during the boom years of Prudhoe Bay production. See "The Facts: Trans Alaska Pipeline System," Alyeska Pipeline Service Company, 2007.

163. The flows of large waterfalls have large seasonal fluctuations and are thus difficult to pin down. The following average flow estimates are from www.world-waterfalls.com. The three numbers are the height h in meters, the flow I expressed as a multiple of 10³ m³/sec, and the power P expressed as a multiple of 10⁹ watts: Inga 96.0 42.4 40.7, Guaira 40.0 13.3 5.32, Niagara 51.0 6.00 3.00, Iguazu 82.0 1.72 1.39, and Victoria 107.0 1.09 1.14. Their relationship is $P = I\rho gh$, where $\rho = 10^3$ kg/m³ is the density of water and g = 9.8 m/sec² is the acceleration due to gravity. Note that Guaira Falls no longer exist. See J. Vidal, "Banks Meet over £40bn Plan to Harness Power of Congo's River and Double Africa's Electricity," The Guardian, 21 Apr 08; B. Box, J. Eddington and M. Day, *Brazil Handbook* (Footprint Handbooks, 2003).

164. The three numbers below have the same meaning as for the waterfalls above: Itaipu 196.0 8.79 16.9, Three Gorges 101.0 15.0 14.9, Grand Coulee 116.0 3.10 3.52, Volga GES 44.0 8.05 3.47, Aswan High 111.0 1.74 1.90, and Hoover 221.0 0.39 0.85. The flows quoted are rough yearly averages. See R. M. L. Ferreira *et al.*, eds., *River Flow 2006* (Taylor and Francis, 2006).

165. The retail value (in U.S. dollars) of the electricity flowing in a 765-kilovolt transmission line is about 1.4 × 10⁹ watts × $0.10/kwh / (1,000 watts/kw × 3,600 sec/h) = $39 per second.

166. The bridge traffic volumes quoted below are half the total, because tolls are collected only in one direction (www.goldengatebridge.org and www.panynj.gov). The three numbers are vehicles per year expressed as a multiple of 10⁷, toll per vehicle in U.S. dollars, and toll total cash flow in U.S. dollars per second: Golden Gate 2.0 6.0 3.85, George Washington 5.2 6.0 9.98, and Verrazano 3.8 10.0 12.00.

167. According to the McDonald's 2008 10-K report submitted to the Securities and Exchange Commission, the company's total retail sales that year were $71 billion, and the total number of restaurants, company-owned and franchised, was just under 32,000. The number of restaurants required to sell hamburgers and milkshakes at

$39 per second was thus $39/sec × 3,600 sec/h × 24 h/d × 365 d/y × 32,000 restaurants / $7.1 × 10^{10}/year = 550 restaurants. However, high-traffic restaurants sell
more than average ones. A commonsense estimate of the number of high-traffic
restaurants required is roughly $39/sec × 60 sec/min / (0.5 sales/min × $5/sale × 4
lines) = 234 restaurants. New York City presently has just over 60 McDonald's
restaurants.

168. The interest on $10 billion borrowed at consumer rates (roughly 10%) is
$1.0 × 10^{10} × 0.1 $year^{-1}$ / (365 d/y × 24 h/d × 3,600 sec/h) = $31.70 per second.

169. See, for example, B. Greenberg, V. P. Bindokas, and J. R. Gauger, "Biological
Effects of a 765-kV Transmission Line: Exposures and Threshold in Honeybee Col-
onies," Bioelectromagnetics **2**, 315 (1981).

170. R. Bartikas and K. D. Srivastava, *Power and Communications Cables:
Theory and Applications* (Wiley-IEEE, 2003). The best sources of technical information about submarine power cables are the websites of companies that manufacture them. See, for example, ABB (www.abb.com), Nexans (www.nexans.com),
Prysmian (www.prysmian.com), and JS Neoplant (www.jsneoplant.com).

171. A 750-kilovolt transmission line with typical size parameters (see note 156)
and also h = 30 meters above the ground produces a fringing field on the ground of
$\mathbf{E} = -V_0$ (0,1,0) Re (a exp(iξ) + b exp(iξ + i2π/3) + c exp(iξ − i2π/3), where a = 2h /
$(x^2 + h^2)$, b = $2h/((x-d)^2 + h^2)$, and c = $2h/((x+d)^2 + h^2)$. Note that this expression
includes the transmission line's mirror image below the ground plane caused by
the highly conductive earth. The peak field strength V_0 ($a^2 + b^2 + c^2$ − ab − bc −
ca)$^{1/2}$ is maximized at x = ±22.3 meters and equals 2.33 × 10^3 volts/meter. The maximum magnetic field strength is 2.33 × 10^3 volts/m / (3.0 × 10^8 m/sec) = 7.77 ×
10^{-6} tesla.

172. A party balloon (radius 15 cm and mass 2.3 g) will stick to the ceiling after
being rubbed on one's hair if its potential exceeds $(mg/\pi\varepsilon_0)^{1/2}$ = (2.3 × 10^{-3} kg × 9.8
m/sec^2 / (π × 8.85 × 10^{-12} farads/m))$^{1/2}$ = 2.85 × 10^4 volts. The electric field generated by such a stuck balloon at the location of one's head (1 meter below the ceiling)
is 4 × 2.85 × 10^4 volts × (0.15 m)^2 / (1 m)^3 = 2.57 × 10^3 volts per meter.

173. Estimates for the construction cost of new 765-kilovolt line range from $1.6
million to $2.5 million per kilometer ($2.5 million to $7.1 million per mile). See E.
O'Grady and C. Wiessner, "OGE Energy Pursued 765-kV Line for Wind Genera-
tion," Reuters, 15 Jul 08; B. Woodall, "2-PJM OKs 2 U.S. Mid-Atlantic Transmission
Lines," Reuters, 22 Jun 07; "Interstate Transmission Vision for Wind Integration,"
Electricity Today, Sep 07, p. 30.

174. T. J. Hammons, G. Palmasson, and S. Thorhallsson, "Geothermal Electric
Power Generation in Iceland for the Proposed Iceland/United Kingdom HVDC
Power Link," IEEE Transaction on Energy Conversion **6**, 289 (1991); J. Leake and
R. Booth, "Iceland's Hot Rocks May be Power Source for UK," London Times, 13
May 07.

175. M. G. Morris, "Building the Next Interstate System," Public Utilities Fortnightly, Jan 06; A. Stone, "Feds to Take Control of Electric Superhighway," Forbes, 9 Mar 09; "AEP Signs on as Corporate Sponsor of Pickens' Energy Plan," Reuters, 19 Feb 09; P. Fairley, "Building an Interstate Highway System for Energy," Discover Magazine, 10 Jun 09. See also P. M. Grant, C. Starr, and T. J. Overbye, "A Power Grid for the Hydrogen Economy," Scientific American, 26 Jun 06.

176. The peak current carried in each wire of the 765-kilovolt power line is $2\pi\varepsilon_0 c$ $V_0 = 2\pi \times 8.85 \times 10^{-12}$ farads/m $\times 3.0 \times 10^8$ m/sec $\times 8.96 \times 10^4$ volts $= 1.49 \times 10^3$ amperes. The resistance per unit length for the (aluminum) wire is (see notes 155, 156) 2.82×10^{-8} ohm m / $(\pi \times (0.015 \text{ m})^2) = 4.0 \times 10^{-5}$ ohms/m. The skin effect and steel core of the wire increase this number only by about 10%. The average power dissipated per meter is thus $0.5 \times (1.49 \times 10^3 \text{ amp})^2 \times 4.0 \times 10^{-5}$ ohms/m $=$ 44.4 watts/m. For a three-wire transmission line 1,000 kilometers in length, the fraction of the total power lost to resistive heating is 3×44.4 watts/m $\times 10^6$ m / $(1.4 \times 10^9 \text{ watts}) = 0.09$. An adult human resting in a chair generates about 100 watts of heat.

177. According to the Department of Energy's "International Energy Annual" (www.eia.doe.gov/iea), 66% of the world's 2006 electricity generation came from conventional thermal sources (i.e., from coal and natural gas); 17% came from hydroelectric sources; 15% came from nuclear sources; and 2% came from all other sources, including wind, solar, and biomass.

178. This is unfortunately a sensitive subject because building transmission lines with government blessing (and subsidies) requires identifying the lines as green. There is not much discussion of coal-by-wire, the idea of siting power plants near coal mines and piping the electricity they generate to cities far away, even though some of the energy superhighway amounts to this. See L. Corum, "Creating the 21st Century U.S. Grid," Renewable Energy Focus **10**, 40 (2009); S. Greenwald and J. Gray, "Transmission Superhighway or Interconnected Patchwork?," Power Magazine, 1 Apr 09; D. Sassoon, "Transmission Superhighway on Track to Carry Cheap, Dirty Coal Power to Northeast," SolveClimate, 24 Mar 09.

179. The factor of 1.5 between the minimum electric power demand and the maximum at the afternoon rush hour may be seen in outlook graphs published daily by the California Independent System Operator (Cal ISO) at www.caiso.com. The time-dependent demand is given roughly by $P(t) = P_0 (1 + 0.2 \cos(2\pi t/T))$, where $T = 24$ hours. Assuming that the ratio is roughly the same everywhere, then New York City's maximum demand exceeds its minimum demand by $0.4 \times 6.28 \times 10^9$ watts $= 2.51 \times 10^9$ watts (see note 154).

180. The amount of energy contained in 1,000 kilometers of 765-kilovolt transmission line is 1.4×10^9 watts $\times 10^6$ m / $(3 \times 10^8$ m/sec$) = 4.67 \times 10^6$ joules (see note 156). A generic red wine glass (10 fluid ounces) of gasoline holds 10 oz $\times 719$ kg/m^3 $\times 4.4 \times 10^7$ joules/kg / $(3.38 \times 10^4$ oz/m$^3) = 9.36 \times 10^6$ joules. For comparison, a

Big Mac meal with supersize fries has an energy content of 1,230 kilocalories $\times 10^3$ cals/kcal $\times$ 4.19 joules/cal = 5.15×10^6 joules.

181. The time it takes to drain a 1,000-kilometer transmission line is 10^6 m / 3 $\times$ 10^8 m/sec = 3.3×10^{-3} sec. Nerve propagation speed depends greatly on the type of nerve in question. Assuming the very fast value of 100 meters per second appropriate for myelinated fibers, the time it takes a nerve impulse to travel from your fingertip to your brain is 1 m / (100 m/sec) = 0.01 sec. See A. G. Brown, *Nerve Cells and Nervous Systems: An Introduction to Neuroscience* (Springer, 2001).

182. The atomic unit of electric field is R_∞ / (ea$_0$) = 2.57×10^{11} volts/meter (see notes 107 and 109). Here are some avalanche breakdown fields (dielectric strengths) of insulators expressed in multiples of 10^7 volts per meter: Air 0.3, Corning 7900 Glass 0.35, Strontium Titanate 0.8, Paper 1.6, Porcelain 1.7, Polyethylene 2.0, Fused Quartz 4.0, CPVC (Chlorinated Polyvinyl Chloride) 4.9, Tefzel (Ethylene Tetrafluoroethylene) 5.0, Mica 18.0, TecaMax (Polyphenylene) 25.0. These values have large error bars that are difficult to quantify, especially for values above 10^8 volts per meter. See www.boedeker.com; V. Shah, *Handbook of Plastics Testing and Failure Analysis* (Wiley-Interscience, 2007); C. A. Harper, *Handbook of Ceramics, Glasses and Diamonds* (McGraw-Hill, 2001); R. L. Freeman, *Fundamentals of Telecommunications* (Wiley-IEEE, 2005).

183. The maximum electric field energy density allowed by dielectric strength of insulators is roughly $0.5 \times 8.85 \times 10^{-12}$ farads/m $\times$ (10^8 volts/m)^2 = 4.43×10^4 joules/m^3 (see note 182). The energy density of gasoline is greater than this by the factor 4.4×10^7 joules/kg $\times$ 719 kg/m^3 / 4.43×10^4 joules/m^3 = 7.14×10^5.

184. No one knows for sure whether all materials become insulators or superconductors when they're cooled sufficiently, because it isn't technically possible to refrigerate things to absolute zero. Metals that don't become superconducting, even at the lowest temperatures achieved thus far, include Cu, Ag, and Au; the alkali metals; the alkaline earth metals; and the ferromagnetic metals Fe, Co, and Ni. Some of these become superconducting under pressure, but this does not count because theory predicts that everything will become superconducting at sufficiently high pressure.

185. The magnetic field with the same energy density as the avalanche-limit electric field is 10^8 volts/m / (3×10^8 m/sec) = 0.33 tesla (see note 183). This is roughly the highest field that can be sustained by pure (type-I) superconductors. (A magnetic field of 0.3 tesla destroys the superconductivity of elemental tantalum, a famous borderline case.) This is thus the largest field in which energy can be stored with no loss. Much higher fields can be sustained by superconductors penetrated by magnetic vortices (type-II), but these exhibit some loss due to vortex motion. The commercial type-II superconductor Nb-Sn can tolerate fields as high as 30 tesla, and small commercial magnets made from it routinely deliver fields of 20 tesla. Gasoline's energy density exceeds that of the 30 tesla field by the factor $7.15 \times 10^5 \times$ (0.33 tesla / 30.0 tesla)^2 = 87. The cuprate superconductors,

such as $Bi_2Sr_2Ca_2Cu_3O_7$, can probably tolerate fields over 100 tesla, but nobody knows the exact number because it's too high to measure with existing technology. The cuprates are unfortunately difficult to use for power applications because they are brittle and have poor vortex pinning. For comparison, the strongest known permanent ferromagnet, $Nd_2Fe_{14}B$, generates a magnetic field of 1.4 tesla. See B. T. Matthias, T. H. Geballe, and V. B. Compton, "Superconductivity," Rev. Mod. Phys. **35**, 1 (1963); R. D. Parks, *Superconductivity* (CRC Press, 1969); D. C. Jiles, *Introduction to Magnetism and Magnetic Materials* (CRDC Press, 1998); M. B. Maple *et al.*, "Unconventional Superconductivity in Novel Materials," in *Supercon-ductivity: Conventional and Unconventional Superconductors, Vol. 1*, K. H. Ben-neman and J. B. Ketterson, eds. (Springer, 2008); Y. Ando *et al.*, "Resistive Upper Critical Fields and Irreversibility Lines of Optimally Doped High-T_c Cuprates," Phys. Rev. B **60**, 12475 (1999).

186. The main bending magnets at the Large Hadron Collider in Geneva are made with a Nb-Ti alloy that can tolerate fields as high as 15 tesla and can operate comfortably at 8.33 tesla. Each magnet is 14.5 meters long and has two beamline bores with magnetic radius 0.057 meter. The stored energy per magnet is thus roughly 8.85×10^{-12} farads/m $\times \pi \times (0.058$ m $\times 3 \times 10^8$ m/sec $\times 8.33$ tesla$)^2 \times 14.5$ m $= 8.47 \times 10^6$ joules. The official storage capacity is 7.4×10^6 joules (3.5 sticks of dynamite) (see note 113). The total energy stored in the accelerator is thus 7.4×10^6 joules/magnet $\times 1,232$ magnets $= 9.12 \times 10^9$ joules. See A. Bézaguet *et al.*, "The Superfluid Helium Model Cryoloop for the CERN Large Hadron Collider (LHC)," in *Advances in Cryogenic Engineering, Vol. 39a*, P. Kittel, ed. (Springer, 1994); N. Siegel, "Status of the Large Hadron Collider and Magnet Program," IEEE Trans. Appl. Supercond. **7**, 252 (1997); R. Perin and D. Leroy, "Superconducting Magnets for Particle Accelerators (Dipoles, Multipoles)," in *Handbook of Applied Supercon-ductivity*, R. Perin and D. Leroy, eds. (Taylor and Francis, 1998); D. Overbye, "Giant Particle Collider Struggles," New York Times, 3 Aug 09.

187. The Boeing 747-400 has a kinetic energy at cruise of $0.5 \times 4.0 \times 10^5$ kg $\times (253$ m/sec$)^2 = 1.28 \times 10^{10}$ joules, not including the weight loss due to fuel burn at takeoff. A big transmission line delivers this much energy in 1.28×10^{10} joules / $(1.4 \times 10^9$ watts$) = 9.1$ seconds. The energy required to lift the plane to cruising al-titude is 4.0×10^5 kg $\times 9.8$ m/sec$^2 \times 35,000$ ft / $(3.28$ ft/m$) = 4.18 \times 10^{10}$ joules. The line delivers this much energy in 30 seconds.

188. At \$0.10 per kilowatt hour, the cost of the electricity required to bring a Boe-ing 747 to cruising speed (at 100% efficiency) is 1.28×10^{10} joules $\times$ \$0.10 kwh^{-1} / $(3.6 \times 10^6$ joules/kwh$) = \$356$. The cost to raise the plane to cruising altitude is \$1,161 (see notes 165 and 187).

189. The amount of energy storage required to even out New York City's demand over twenty-four hours is roughly 2.51×10^9 watts $\times 3,600$ sec/h $\times 12$ h / $\pi = 3.45 \times 10^{13}$ joules (see note 179). The Hiroshima atomic bomb released energy 6.27×10^{13} joules. The daily storage need of New York city is thus 0.55 Hiroshima-sized atomic bombs.

190. According to the U.S. Energy Information Administration's "International Energy Annual (IEA)" (www.eia.doe.gov/iea), the world's average electric power consumption in 2006 was 1.64×10^{16} wh/y / (365 d/y $\times$ 24 h/d) = 1.87×10^{12} watts. The daily storage needs of the world are thus $0.55 \times 1.87 \times 10^{12}$ watts / (6.28×10^9 watts) = 163 Hiroshima-sized atomic bombs (see notes 154, 179, and 189).

191. The U.S. Energy Information Administration's "Electric Power Annual" identifies the total 2007 pumped storage capacity in the United States as 2.2×10^{10} watts. The two numbers below are maximum power expressed as a multiple of 10^9 watts and total storage expressed as a multiple of 10^{13} joules for the largest U.S. facilities: Bath County 2.77 12.9, Ludington 1.87 10.9, Castaic 1.56 68.0, Raccoon Mountain 1.53 15.1, Helms 1.20 70.0, Yards Creek 0.40 1.3, Blenheim Gilboa 1.16 6.8, Northfield 1.08 5.1, Bad Creek 1.07 10.1, Muddy Run 1.07 5.0, Rocky Mountain 0.85 2.3, Jocassee 0.61 12.9, Bear Swamp 0.6 1.3, Fairfield 0.51 2.3, Seneca 0.43 2.0, Gianelli 0.42 6.0, Taum Sauk 0.45 1.4, Cabin Creek 0.32 0.5, M. Elbert 0.2 4.0. Note that both numbers come primarily from the facilities' own promotional literature and have uncertainties of 20% or more because of measurement difficulties. The total energy reported is ρghV, where ρ is the density of water, g is the acceleration due to gravity, h is the head, and V is reservoir volume. According to the EIA's "International Energy Annual (IEA)" (www.eia.doe.gov/iea), the average electric power consumption in the United States in 2006 was 3.82×10^{15} wh/y / (24 h/d $\times$ 365 d/y) = 4.36×10^{11} watts. Assuming that the surge power is 1/5 of this total, the fraction of U.S. surge demand potentially handled by pumped storage in 2006 was thus 2.2×10^{10} watts $\times$ 5 / (4.36×10^{11} watts) = 0.25 (see note 179). The situation in the world as a whole is similar. See Task Committee on Pumped Storage, *Hydroelectric Storage Technology: International Experience* (Am. Soc. Civil Eng., 1996).

192. In 1996 dollars, the capital cost per kilowatt for a stand-alone hydropump facility was between $600 and $1,100, while the capital cost per kilowatt for a gas turbine was about $450. Source: T. C. Elliott, K. Chew, and R. C. Swanekamp, *Standard Handbook of Powerplant Engineering* (McGraw-Hill, 1997) (see note 191).

193. A notorious case of this problem is the nuclear electricity industry in France, which supplies 80% of its power, but not, as it turns out, 80% of its surge capacity. See M. Schneider, "Nuclear Power in France—Beyond the Myth," Greens-EFA Group in the European Parliament, December 2008. According to Schneider, France handles some of its surge problem by running its reactors in load-following mode, which normally isn't done in other countries because it loses money, and then handles the rest by importing peak power from neighboring countries and exporting back baseline capacity at a loss.

194. Storage of compressed air underground has the specific weakness of working only in certain geological formations. See S. Patel, "The Return of Compressed Air Storage," Power Magazine, 15 Oct 09; W. F. Pickard, N. J. Hansing, and A. C. Shen, "Can Large-Scale Advanced-Adiabatic Compressed Air Energy Storage Be

Justified Economically in an Age of Sustainable Energy?," J. Renew. Sustain. Energy **1**, 033102 (2002).

195. Estimated efficiencies of hydropump facilities range from 70 to 86% but are typically between 75 and 80%. See P. A. Breeze, *Power Generation Technologies* (Newnes, 2005); W. Shepherd and D. W. Shepherd, *Energy Studies* (Word Sci., 2003). Estimated efficiencies of coal-fired steam turbines range from 31 to 55%, but are typically between 35 and 39%. See F. Kreith, *The CRC Handbook of Mechanical Engineering* (CRC Press, 1998).

196. The world's nuclear weapons stockpile is considerably smaller than it was at the height of the Cold War, but the precise numbers remain classified. According to documents released by U.S. Energy Secretary Hazel O'Leary in a press conference on June 27, 1994, the U.S. stockpile had decreased by 1994 to 2,375 megatons from its 1960 peak of just over 20,000 megatons. These documents are filed at the U.S. Office of Scientific and Technical Information (https://www.osti.gov/opennet/press.jsp) under "Documents Provided at the Secretary's June 1994 Press Conference." Assuming that the world never had more explosive power than 40,000 megatons, the total energy in question is less than 4.0×10^{10} tons of TNT $\times 4.2 \times 10^9$ joules/ton of TNT $= 1.68 \times 10^{20}$ joules. This is roughly the world's consumption of oil (or natural gas) in one year (see note 5).

197. Precise numbers about the maintenance costs of hydroelectric machinery are difficult pin down even though various sources describe them as "very low." See, for example, S. R. Brockschink, J. H. Gurney, and D. B. Seely, "Hydroelectric Power Generation," in *The Electric Power Generating Handbook*, Grigsby, ed. (CRC Press, 2000). According to the U.S. Energy Information Administration's "Electric Power Annual," maintenance costs for U.S. hydroelectric facilities in 2009 were $\$3.5 \times 10^{-3}$ kwh^{-1} and thus accounted for about 3.5% of the electricity's retail cost.

198. The area of Lake Mead, when full, is 640 square kilometers. The height of Hoover Dam is 221 meters. The energy stored in the top meter of Lake Mead is thus 6.4×10^8 m$^2 \times 1$ m $\times 1,000$ kg/m$^3 \times 9.8$ m/sec$^2 \times 221$ m $= 1.39 \times 10^{15}$ joules. This is 40.3 times the surge requirement of New York City, 0.58 times the surge requirement of the United States, and 0.13 times the surge requirement of the world (see notes 189, 190, and 191).

199. Modeling Lake Mead as an inverted pyramid gives a total stored energy of $(221$ m $/ 1$ m$) \times 1.39 \times 10^{15}$ joules $/ 4 = 7.68 \times 10^{16}$ joules (see note 198). The number of Lake Meads required to supply the world with electric energy for one month is thus $1/12 \times 1.64 \times 10^{13}$ kwh $\times 3.6 \times 10^6$ joules/kwh $/ (7.68 \times 10^{16}$ joules$) = 64$ (see note 190).

200. The area of sixty-four Lake Meads is (see note 198) $64 \times 6.4 \times 10^8$ m$^2 / (1.97 \times 10^{12}$ m$^2) = 0.021$ times the area of Mexico and 0.007 times the area of Canada or the Sahara (see note 46).

201. The huge Dinorwig and Ffestiniog facilities in northern Wales provide a sobering example of the competitive difficulties of hydropump storage vis-à-vis

natural gas peaking capacity. The Dinorwig facility was built for the specific purpose of balancing loads for nuclear plants. Its huge capacity (1.8×10^9 watts for 5
hours) took ten years to build and had an official cost of £450 million ($2.99 billion
in 1995 dollars, assuming the figures are from 1975). The Ffestiniog facility had an
official cost of £103 ($263 million in 1995 dollars, assuming the figures are from
1970). However, as part of Britain's national grid privatization, the two facilities
were sold together in 1995 to Edison International for £650 million ($1.03 billion
in 1995 dollars). They are now jointly owned by International Power plc (75%) and
Mitsui and Co., Ltd. (25%), a huge Japanese conglomerate. Meanwhile, the fraction
of U.K. electricity generation supplied by natural gas went from 0% in the early
1990s (when it was banned) to 38% in 2003. See I. P. Burdon, "Gas and the Current
Generation Mix in England and Wales," Power Engineering Journal **14**, 61 (2000);
P. Brown, "A Vision of Britain in 2020: Power Cuts and the 3-Day Week," The
Guardian, 1 Jul 03; P. Rodgers, "Windfall Boost for Grid Directors," The Indepen-
dent, 29 Dec 95; J. Chesshire, "U.K. Electricity Supply under Public Ownership,"
in *The British Electricity Experiment—Privatisation: The Record, the Issues, the
Lessons*, J. Surrey, ed. (Earthscan, 1996); E. Williams, *Dinorwig—The Electric Moun-
tain* (U.K. Central Generation Board, National Grid Division, 2000); C. Middleton,
"Wales: Here Be Power Dragons," Daily Telegraph, 20 Nov 00; www.fhc.co.uk.

CHAPTER 6: INSPIRING MAMMOTHS

202. The Ford Nucleon was a 1958 concept car powered by a small nuclear re-
actor. See M. Doeden, *Crazy Cars* (Lerner Classroom, 2007); "Model of Atom Car
Displayed by Ford," New York Times, 14 Feb 58; "Vacuum Cleaners Eyeing the
Atom," New York Times, 11 Jun 55; K. Rose, *One Nation Underground: The Fallout
Shelter in American Culture* (New York University Press, 2004); C. West, *Fallout
Shelter Handbook* (Fawcett, 1962).

203. Ongoing antinuclear (and anticoal) political activity may be found at the
Nuclear Information and Resource Service (www.nirs.org). A. B. Lovins and J. H.
Price, *Non-Nuclear Futures: The Case for an Ethical Energy Strategy* (Harper-
Collins, 1980); B. Smith, *Insurmountable Risks: The Dangers of Using Nuclear
Power to Combat Global Climate Change* (RDR Books, 2006); J. Rifkin, "No Nukes!"
Los Angeles Times, 29 Sep 06; H. Caldecott, *Nuclear Power Is Not the Answer* (New
Press, 2006).

204. There are intense political struggles over nuclear power in all industrialized
countries at the moment. See C. Joppke, *Mobilizing Against Nuclear Energy: A Comparison of Germany and the United States* (U. Calif. Press, 1993); S. Dolling, "The
Renaissance of the Anti-Nuclear Movement," Der Spiegel, 10 Nov 08; E. Arita,
"Rokkasho Plant Too Dangerous, Costly: Expert," Japan Times, 27 Nov 08; M.
Brunswick, "Minnesota House Says No to New Nuclear Power Plants," Minneapo-
lis-St. Paul Star Tribune, 30 Apr 09; K. Garber, "Gauging the Prospects for Nuclear

Power in the Obama Era," U.S. News and World Report, 27 Mar 09; B. Bender, "Obama Seeks Global Uranium Fuel Bank," Boston Globe, 8 Jun 09; T. Czuczka, "Merkel Retreats on Nuclear Power, Backs Electric Cars," Bloomberg, 29 Jun 09; R. Nolan, "Nuclear-Power Debate Reignites in Germany," Time, 9 Jul 09; M. L. Wald, "Sole U.S.-Owner Reactor Fuel Plant Denied Loan," New York Times, 28 Jul 09; W. Boston, "Japan's Nuclear Crisis Has Germany's Merkel Worried," Time, 21 Mar 11.

205. An important U.S. pronuclear group is the Clean and Safe Energy Coalition, co-chaired by former New Jersey governor Christine Todd Whitman and former Greenpeace activist Patrick Moore (www.cleansafeenergy.org). See G. P. Zachary, "The Case for Nuclear Power," San Francisco Chronicle, 5 Feb 06; P. Moore, "Going Nuclear," Washington Post, 16 Apr 06; E. Marris, "Nuclear Reincarnation," Nature **441**, 796 (2006); E. Rosenthal, "Italy Embraces Nuclear Power," New York Times, 23 May 08; C. T. Whitman, "California Should Revive Nuclear Energy Option," Sacramento Bee, 3 Aug 09. Much of this thinking was anticipated in H. A. Bethe, "The Necessity of Fission Power," Scientific American **234**, No. 1, 21 (1976).

206. S. Peterson, "Asia Hungry for Nuclear Power," Christian Science Monitor, 30 Jun 04; A. E. Cha, "China Embraces Nuclear Future," Washington Post, 29 May 07; K. Hall, "Toshiba-Westinghouse Takes China Nuclear," Business Week, 18 Dec 06. The two other highly pro-nuclear countries are Japan and France. According to the BP Statistical Review of World Energy 2008, nuclear energy presently provides 8.1% of the energy consumed in the United States, 12% of the energy consumed in Japan, 14% of the energy consumed in South Korea, and 39% of the energy consumed in France. See E. Cue, "How France Sees Its Nuclear-Powered Future," U.S. News and World Report, 10 Mar 09.

207. The radioactivity issues of spent fuel from light-water reactors are thoroughly surveyed in the U.S. National Research Council publication, "Nuclear Wastes: Technologies for Separations and Transmutation" (National Academies Press, 1996). On time scales shorter than one thousand years, danger comes entirely from fission products, chiefly ^{90}Sr and ^{137}Cs. These are very hot, but their activity decays in a few hundred years to a value below that of the ^{238}U in which they're embedded. ^{238}U is not dangerous (except chemically). Beyond one thousand years, the chief danger is from (1) transuranic elements (Pu, Am, Np) created in the reactor from ^{238}U by neutron capture, and (2) the long-lived fission products ^{99}Tc and ^{129}I. These substances can, in principle, be extracted chemically and recycled into the reactor, where they burn or neutralize by neutron capture. However, with the exception of Pu, which several countries routinely extract for fuel, such remediation isn't done anywhere in the world at the moment, for cost reasons. The waste thus remains a health threat for much longer than one thousand years. The seriousness of this threat is not known. The Chernobyl "experiment" has suggested that it might be less than everyone thought, and some people even argue that there is no such problem. See W. Tucker, "There Is No Such Thing as Nuclear Waste," Wall Street Journal, 13 Mar 09.

208. This is an extremely sensitive matter, and most countries represent that they do have nuclear waste storage facilities, just not the final ones. There is also sensitivity about the amounts of radioactivity in the various kinds of waste, some of which needs to be stored for geologic time and some of which doesn't. But there is always some component of used nuclear fuel that requires removal from the biosphere and contact with groundwater for thousands of years, and no country has committed itself to a long-term solution for storing this component. See M. L. Wald, "Is There a Place for Nuclear Waste?," Scientific American, 3 Aug 09.

209. Civilian nuclear power is usually seen as beginning with the third commercial power plant in the United States, the Yankee-Rowe, which was built in 1960 and began service in 1961. See A. Gorlick, "Yankee Plant Closed but Its Waste Remains," Boston Globe, 21 Aug 07.

210. This is difficult to document thoroughly on account of the need, for antiterrorism reasons, to keep security breaches and black market activity out of the news. See J. Ryan, "Report Urges Focus on Nuclear Terrorism," ABC News, 20 Nov 08; M. Bunn, "The Risk of Nuclear Terrorism—and Next Steps to Reduce the Danger," Testimony for the Committee on Homeland Security and Governmental Affairs, U.S. Senate, 2 Apr 08; M. Levi, *On Nuclear Terrorism* (Harvard Univ. Press, 2009).

211. The extent of state control of nuclear information around the world is difficult to document, but one telltale symptom is the abundance of nuclear protest news in some countries and the complete absence of it in others. See H. Flam, *States and Anti-Nuclear Movements* (U. Edinburgh Press, 1994); "Nuclear Power Protest Leads to Riot in Taiwan," New York Times, 19 Oct 96; C. Brown and M. Harrison, "Greenpeace Protest on Nuclear Energy Forces Blair to Switch Venue," The Independent, 30 Nov 05; E. Shirbon and C. Evans, "Greenpeace Climb Eiffel Tower in Nuclear Protest," Reuters, 13 Jul 08.

212. α particles are ^{4}He nuclei. Heavy radioactive nuclei emit them with kinetic energies between 4 and 10 MeV, but typically 6 MeV $\times$ 1.602 $\times$ 10^{-13} joules/MeV = 9.6 $\times$ 10^{-13} joules, which is about a million times greater than the 4.4 $\times$ 10^7 joules/kg $\times$ 0.014 kg/mole / 6.022 $\times$ 10^{23} atoms/mole = 1.02 $\times$ 10^{-18} joules per atom released when gasoline burns. Their corresponding speed is about $(1 - (1 + 9.61 \times 10^{-13}$ joules / $(6.64 \times 10^{-27}$ kg $\times (3 \times 10^8$ m/sec$)^2)^{-2})^{1/2}$ = 0.06 times the speed of light. β particles are electrons or positrons. Radioactive nuclei emit them with energies between zero and 5 MeV. A 2 MeV $\times$ 1.602 $\times$ 10^{-13} joules/MeV = 3.20 $\times$ 10^{-13} joule β particle has a speed of about $(1 - (1 + 3.20 \times 10^{-13}$ joules / $(9.11 \times 10^{-31}$ kg $\times (3 \times 10^8$ m/sec$)^2)^{-2})^{1/2}$ = 0.98 times the speed of light. γ rays have energies in this same range and travel at the speed of light. See H. A. Bethe and P. Morrison, *Elementary Nuclear Theory*, 2nd ed. (Dover, 2006); G. F. Knoll, *Radiation Detection and Measurement*, 3rd ed. (Wiley, 2000); E. Segré, *Nuclei and Particles* (Benjamin, 1965); E. Fermi, *Nuclear Physics: A Course Given by Enrico Fermi at the University of Chicago* (U. Chicago Press, 1974). See nuclide chart at www-nds.iaea.org.

213. The symptoms of radiation sickness are described in "Acute Radiation Syndrome: A Fact Sheet for Physicians," U. S. Centers for Disease Control and Prevention, 18 Mar 05 (www.bt.cdc.gov). The cellular damage responsible for radiation sickness repairs itself over time, so the dose responsible must be massive and quickly administered. Victims either succumb to the illness (usually because of bone marrow destruction) or recover fully. The amount of radiation required to induce acute illness is roughly one nuclear absorption event per cell. In the specific case of γ rays, which are highly penetrating and thus especially dangerous, the relevant dose is 0.7 joules deposited per kilogram of tissue, or 0.7 gray. (This is the same dosage as 70 rads or 0.7 sieverts.) The cancer-inducing effects of radiation are more difficult to assess because they act in conjunction with other factors and show up only over long times. The population exposed to atomic bomb radiation in Hiroshima and Nagasaki showed a marked increase in leukemia. The population exposed to fallout from the Chernobyl incident didn't, although it showed a marked increase in thyroid cancer, presumably due to ingestion of radioactive iodine from tainted milk. There are no significant birth defect signals in either population. See L. H. Hempelmann and J. G. Hoffman, "Practical Aspects of Radiation Injuries," Annu. Rev. Nucl. Sci. **3**, 369 (1953); L. H. Hempelmann, C. L. Lushbaugh, and G. L. Voelz, "What Has Happened to the Survivors of the Early Los Alamos Nuclear Accidents?," Los Alamos National Laboratory, LA-UR-9-2802, August 1979; G. G. Caldwell *et al.*, "Mortality and Cancer Frequency Among Military Nuclear Test (Smoky) Participants, 1957 through 1979," J. Am. Med. Assn. **250**, 620 (1983); J. D. Boice Jr., "Studies of Atomic Bomb Survivors: Understanding Radiation Effects," J. Am. Med. Assn. **264**, 622 (1990); M. Ichimaru *et al.*, "Atomic Bomb and Leukemia: A Review of Forty-Five Years of Hiroshima and Nagasaki Atomic Bomb Survivors," J. Radiation Research Suppl. **32**, 162 (1991); D. Sali *et al.*, "Cancer Consequences of the Chernobyl Accident in Europe Outside the Former USSR: A Review," Intl. J. Cancer **67**, 343 (1996); D. Grady, "Chernobyl's Voles Live But Mutations Surge," New York Times, 7 May 96; L. N. Astakhova *et al.*, "Chernobyl-Related Thyroid Cancer in Children of Belarus: A Case-Control Study," Radiation Research **150**, 349 (1998); V. N. Gapanovich *et al.*, "Childhood Leukemia in Belarus Before and After the Chernobyl Accident: Continued Follow-Up," Env. Biophys. **40**, 259 (2001); E. Cardis, "Cancer Consequences of the Chernobyl Accident: 20 Years On," J. Radiological Protection **26**, 127 (2006).

214. The ability of X-radiation to cause genetic mutations was discovered in 1927 in experiments done on fruit flies. The "very heavy" radiation doses in these experiments were not quantified but were later estimated as high as 150 roentgens × 0.12 gray/roentgen = 18 gray. See H. J. Muller, "Artificial Transmutation of the Gene," Science **66**, 84 (1927); M. Westergaard, "Man's Responsibility to His Genetic Heritage," Bull. Atomic Scientists **11**, No. 9, 318 (1955).

215. P. Leventhal, "More Nuclear Power Means More Risk," New York Times, 17 May 01.

216. The exposed-core accident at the Three Mile Island plant on March 28, 1979, was the first large nuclear incident in the United States. See M. Stencel, "A Nuclear Nightmare in Pennsylvania," Washington Post, 27 Mar 99; B. D. Ayres Jr., "Three Mile Island: Notes from a Nightmare," New York Times, 16 Apr 79. The runaway reactor event at Chernobyl, near the Ukraine-Belorus border, on April 26, 1986, was vastly more serious and is usually regarded as history's most terrible nuclear accident. See S. Schmemann, "Soviet Announces Nuclear Accident at Electric Plant," New York Times, 28 Apr 86; P. Lewis, "Europe After Chernobyl: Cooler Attitudes Toward Nuclear Power," New York Times, 27 Apr 87; G. Medvedev and A. Sakharov, *The Truth About Chernobyl* (Basic Books, 1991); R. Stone, "The Long Shadow of Chernobyl," National Geographic, April 2006. See also W. McKeown, *Idaho Falls: The Untold Story of America's First Nuclear Accident* (ECW Press, 2003). The Fukushima Daiichi accident is still unfolding, but it appears to be the same kind of failure as Three Mile Island except more severe. It is presently the world's second-worst nuclear accident.

217. The subject of stellar nucleosynthesis has a history stretching back to the 1930s, when nuclear physics was just beginning to be understood. See E. M. Burbidge *et al.*, "Synthesis of the Elements in Stars," Rev. Mod. Phys. **29**, 547 (1957); G. Wallerstein *et al.*, "Synthesis of the Elements in Stars: Forty Years of Progress," Rev. Mod. Phys. **69**, 995 (1997); S. E. Woosley, A. Heger and T. A. Weaver, "The Evolution and Explosion of Massive Stars," Rev. Mod. Phys. **74**, 1015 (2002); D. D. Clayton and L. R. Littler, "Astrophysics with Presolar Stardust," Annu. Rev. Astron. Astrophys. **42**, 39 (2004).

218. S. Fetter, "How Long Will the World's Supply of Uranium Last?," Scientific American, 9 Mar 09. The estimate of 230 years in this paper is based on current use patterns and assessments of the total world uranium supply from the "Red Book," International Atomic Energy Agency Publication NEA-6345, available as *Uranium 2007: Resources, Production and Demand* (OECD Press, 2008). According to this report, (1) the world presently uses about 7.0×10^4 tonnes of uranium metal per year, (2) the world has a "reasonably assured and inferred" uranium supply of about 5.5×10^6 tonnes, and (3) the world has a likely "undiscovered" uranium supply of 10.5×10^6 tonnes. With the open-cycle practices used in most of the world, including the United States, the world's uranium supply would thus last for $(5.5 + 10.5) \times 10^6$ tonnes / $(7.0 \times 10^4$ tonnes/year) = 229 years. With a generous 30% from plutonium recycling and another 30% for improved reactor design, the duration would still only be $1.3 \times 1.3 \times 229$ y = 387 years. Nuclear power currently supplies only 15% of the world's electricity. If it were tasked to supply all of the world's electricity it would last at most 0.15×387 years = 58 years. If it were tasked to supply the world's energy of all kinds it would last for 58 y $\times 5.90 \times 10^{19}$ joules/y / $(4.4 \times 10^{20}$ joules/y) = 7.8 years (see notes 5, 177, and 190).

219. Uranium fission yields 215 MeV/atom $\times 1.602 \times 10^{-13}$ joules/MeV = 3.44×10^{-11} joules per atom. The energy content of pure ^{235}U metal is thus 3.44×10^{-11} joules/atom $\times 6.022 \times 10^{23}$ atoms/mole / (0.235 kg/mole) = 8.82×10^{13} joules/kg.

^{235}U comprises only 0.7% of natural uranium. Assuming that the world has 1.6 × 10^7 tonnes of the latter, the total energy available from ^{235}U fission is thus 0.007 × 1.6 × 10^7 tonnes × 10^3 kg/tonne × 8.82 × 10^{13} joules/kg = 9.9 × 10^{21} joules (see note 218). According to the BP Statistical Review of World Energy 2008, the world's remaining coal supply is about 8.5 × 10^{11} tonnes (see note 2). It thus contains energy 8.5 × 10^{11} tonnes × 10^3 kg/tonne × 3.5 × 10^7 joules/kg = 2.98 × 10^{22} joules. The amount of coal energy in the ground thus exceeds the amount of ^{235}U fission energy in the ground by the factor 3.01. With the above generous plutonium recycling and reactor design assumptions, this number reduces to 3.01 / (1.3 × 1.3) = 1.78.

220. Breeding uses capture of fast neutrons to burn nearly all the uranium in the fuel, not just the rare isotope ^{235}U. The maximum theoretical extraction efficiency is about 75% because not all ^{238}U neutron captures produce plutonium and not all the plutonium produced is fissionable. At current burn rates, breeding would thus make the world's uranium supply last for 229 years × 0.75 / 0.007 = 2.45 × 10^3 years (see note 218). However, breeding would also facilitate burning of the world's thorium, which, by most accounts, is four times more abundant than the world's uranium, at least that in the ground. At current burn rates, this would enable the world's fissile fuel to last for 5 × 2.45 × 10^4 years = 1.23 × 10^5 years. However, if the burn rate grew to supply all the world's energy, the fissile fuel would last only 1.23 × 10^5 years × 6.6 y / 387 y = 2.10 × 10^3 years. See R. Wilson, "The Changing Need for a Breeder Reactor," Nuclear Energy **39**, 113 (2000); "Thorium Fuel Cycle—Potential Benefits and Challenges," International Atomic Energy Agency, IAEA-TECDOC-1450, May 2005; U. Schwarz-Schampera, "The Potential of Thorium Deposits," in *Uranium, Mining and Hydrogeology*, B. J. Merkel and A. Hasche-Berger, eds. (Springer, 2008).

221. Information about breeder reactor facilities is highly classified, but the engineering difficulties can be inferred from refereed journal articles and occasional newspaper reports. J. G. Fuller, *We Almost Lost Detroit* (Berkley, 1984); C. K. Matthews, "Liquid Sodium—The Heat Transport Medium in Fast Breeder Reactors," Bull. Mat. Sci. **16**, 477 (1993); P. Hadfield, "Sodium Leak Shuts Japanese Reactor," New Scientist, 16 Dec 95; U. Wehmann, H. Kinjo, and T. Kageyama, "Studies on Plutonium Burning in the Prototype Fast Breeder Reactor Monju," Nuclear Science and Engineering **140**, 205 (2002); F. M. Mitenkov, "Prospects for the Development of Fast Breeder Reactors," Atomic Energy **92**, 453 (2002); V. M. Poplavskii *et al.*, "Danger of Burning of Sodium Coolant," Atomic Energy **96**, 327 (2004); M. Ramesh, "Fast Breeder Reactor Projects Put on Fast Track," The Hindu Business Line, 13 Aug 04; V. I. Kostin and B. A. Vasil'ev, "Problems of BN-800 Construction and the Possibilities of Developing Advanced Fast Reactors," Atomic Energy **102**, 19 (2007); A. Kumar and M. V. Ramana, "Compromising Safety: Design Choices and Severe Accident Possibilities in India's Prototype Fast Breeder Reactor," Science and Global Security **16**, 87 (2008); T. S. Subramanian, "Prototype Fast Breeder Reactor's Massive Safety Vessel Installed," The Hindu, 25 Jun 08; O. Tsukimori, "Japan Fast-Breeder Reactor May Restart in February," Reuters, 12 Aug 09.

222. The French breeder reactor Superphénix, the largest ever built, was attacked by five terrorist rockets while it was under construction in 1982. On May 8, 2008, Green Party activist (and later Swiss cantonal official) Chaïm Nissim told the press that he had carried out the attack. He claimed to have obtained the rockets from the notorious Venezuelan revolutionary Carlos the Jackal. See E. Marshall, "Super Phenix Unscathed in Rocket Attack," Science **215**, 641 (1982); S. Besson, "Après Vingt Ans de Silence, un Ex-Député Avoue l'Attaque à la Roquette Contre Creys-Malville," Le Temps, 8 Aug 03.

223. The plutonium problem of breeding is mainly a matter of quantity, for exactly the same issues come up in fuel rod reprocessing. See "Keep Rein on Plutonium," Los Angeles Times, 27 Jun 98; K. Z. Morgan and K. M. Peterson, *The Angry Genie: One Man's Walk through the Nuclear Age* (U. Oklahoma Press, 1999); F. N. von Hippel, "Plutonium and Reprocessing of Spent Nuclear Fuel," Science **293**, 2397 (2001); J. Bernstein, *Plutonium: A History of the World's Most Dangerous Element* (Cornell U. Press, 2009); A. B. Lovins, "Thorium Cycles and Proliferation," Bull. Atomic Scientists **35**, 16 (1979).

224. "Policy: Putting Brakes on the Fast Breeder," Time, 18 Apr 77; P. David, "Clinch River Fast Breeder: The Final Blow?," Nature **303**, 747 (1983); J. Boudreau, "The American Breeder Reactor Program Gets a Second Chance," Los Alamos Science **2**, 118 (1981); G. Lean, "Want Fast Breeders? Get Rabbits," The Independent, 15 Jun 97.

225. All present-day nuclear reactors breed plutonium. Conventional light-water reactors have breeding ratios (fissionable atoms generated per fission event) of between 0.4 and 0.6. The target breeding ratio of modern fast breeder designs is 1.2. The breeding ratio actually achieved by Superphénix is classified, but most reports place it between 1.06 and 1.16. Some proposed gas-cooled breeder designs have theoretical breeding ratios approaching 1.8. See Y. Ishiwara, Y. Oka and S. Koshizuka, "Breeding Ratio Analysis of a Fast Reactor Cooled by Supercritical Light Water," J. Nucl. Sci. Technol. **38**, 703 (2001); M. Kambe, "Conceptual Design of a Modular Island Core Fast Breeder Reactor 'RAPID-M'," J. Nucl. Sci. Technol. **39**, 1169 (2002); R. L. Garwin and G. Charpak, *Megawatts and Megatons: The Future of Nuclear Power and Nuclear Weapons* (U. Chicago Press, 2002); W. M. Stacey, *Nuclear Reactor Physics* (Wiley-VCH, 2007).

226. In seawater (specific gravity 1.035) the ratio of U ion concentration to Na ion concentration is $(3.4 \times 10^{-6} \text{ kg/m}^3 / 3.02 \text{ kg/m}^3) \times (23.0 + 35.5) / 238.0 = 2.77 \times 10^{-7}$. See B. N. Laskorin, S. S. Metal'nikov, and G. I. Smolina, "Extraction of Natural Uranium from Natural Seawater," Atomic Energy **43**, 1122 (1977).

227. R. Khamizov, D. M. Muraviev, and A. Warshawsky, "Recovery of Valuable Mineral Components from Seawater by Ion-Exchange and Sorption Methods," in *Ion Exchange and Solvent Extraction: A Series of Advances, Vol. 12,* J. A. Marinsky and Y. Marcus, eds. (CRC Press, 1995); M. Kanno, "Present Status of Study on Extraction of Uranium from Seawater," J. Nucl. Sci. Technol. **21**, 1 (1984); K. Saito *et*

al., "Recovery of Uranium from Seawater Using Amidoxime Hollow Fibers," Am. Inst. Chem. Eng. J. **34**, 411 (1988); K. Sekiguchi *et al.*, "Uranium Uptake During Permeation of Seawater Through Amidoxime-Group-Immobilized Micropores," Reactive Polymers **23**, 141 (1994).

228. B. L. Cohen, "Breeder Reactors: A Renewable Energy Source," Am. J. Phys. **51**, 75 (1983).

229. This estimate is deliberately conservative. It is less than Cohen's 7 million years mainly because it assumes that all the world's consumed energy, not just the amount delivered as electricity, is supplied by uranium (see note 228). The oceans contain 1.3×10^{18} m^3 $\times 3.3 \times 10^{-6}$ kg/m^3 = 4.29×10^{12} kg of uranium (see note 226). With breeding, the energy content of this uranium is 4.29×10^{12} kg $\times 8.82 \times 10^{13}$ joules/kg = 3.78×20^{26} joules (see note 219). The uranium in the sea could thus power the world at its present consumption rate for 3.78×10^{26} joules / (4.4×10^{20} joules/y) = 8.6×10^5 years (see note 5).

230. It's thirty billion, actually. This calculation is, however, even more academic than the uranium one, because the relevant fusion technologies don't yet exist. About 0.015% of the hydrogen atoms on earth are deuterium. The reactions that turn deuterium into energy are ^{2}H + ^{2}H $\rightarrow$ ^{3}H + p + 4.03 MeV and ^{2}H + ^{3}H $\rightarrow$ ^{4}He + n + 17.6 MeV. The amount of pure fusion energy one can potentially get out of a cubic meter of seawater is thus $2/3 \times (4.03 + 17.6)$ MeV $\times 1.602 \times 10^{-13}$ joules/MeV $\times 1.54 \times 10^{-4} \times 6.022 \times 10^{23}$ mole^{-1} $\times 10^3$ kg/m^3 / (0.018 kg/mole) = 1.2×10^{13} joules/m^3. The amount of fusion energy we can potentially get out of the sea is therefore 1.3×10^{18} m^3 $\times 1.2 \times 10^{13}$ joules/m^3 = 1.56×10^{31} joules. The fusion energy in the oceans would thus supply the world's energy needs for 1.56×10^{31} joules / (4.4×10^{20} joules/y) = 3.5×10^{10} years (see note 5).

231. The world's first thermonuclear explosion, code named Ivy Mike, consisted of a tank of deuterium lit on fire by a conventional fission bomb. Its 10.4 megaton energy release, which was much greater than expected, obliterated the Pacific island of Elugelab and left behind a crater two kilometers across and thirty meters deep. See C. Seife, *Sun in a Bottle* (Viking Penguin, 2008).

232. J. Nuckolls *et al.*, "Laser Compression of Matter to Super-High Densities: Thermonuclear (CTR) Applications," Nature **239**, 139 (1972); R. C. Arnold, "Heavy-Ion Beam Inertial-Confinement Fusion," Nature **276**, 19 (1978); R. E. Kidder, "Laser-Driven Isentropic Hollow-Shell Implosions: The Problem of Ignition," Nucl. Fusion **19**, 223 (1979); J. J. Duderstadt and G. A. Moses, *Inertial Confinement Fusion* (Wiley, 1982); A. R. Piriz, "Conditions for the Ignition of Imploding Spherical Shell Targets," Nucl. Fusion **36**, 1395 (1996); C. Yamanaka, *Introduction to Laser Fusion* (Routledge, 1991); K. A. Brueckner, ed., *Inertial Confinement Fusion* (Springer, 1993); S. Pfalzner, *An Introduction to Inertial Confinement Fusion* (Taylor and Francis, 2006); J. Randerson, "Lasers Point Way to Clean Energy," The Guardian, 6 Dec 07; I. Sample, "California Fires Up Laser Fusion Machine," The Guardian, 28 May 09.

233. C. J. Hansen, S. D. Kawaler and V. Trimble, *Stellar Interiors* (Springer, 2004).

234. The National Ignition Facility is designed to deliver 1.8×10^6 joules to the capsule hohlraum walls in about 1.0×10^{-8} seconds. The energy in a stick of dynamite is 2.1×10^6 joules (see note 113). The pulse length is intentionally long (and also shaped) to minimize shock wave formation. See C. Bibeau, P. J. Wegner and R. Hawley-Fedder, "UV SOURCES: World's Largest Laser to Generate Powerful Ultraviolet Beams," Laser Focus World **42**, No. 6, 113 (2006); J. Mandel, "National Ignition Facility Prepares for Fusion Test," Scientific American, 20 Aug 09; D. Lyons, "Could This Lump Power the Planet?," Newsweek, 13 Nov 09.

235. The NIF laser delivers average power 1.8×10^6 joules / $(1.0 \times 10^{-8}$ sec$) = 1.8 \times 10^{14}$ watts to its target (see note 234). It delivers peak power 5.0×10^{14} watts $\times$ 3,600 sec/h $\times$ 24 h/d $\times$ 365 d/y / $(4.4 \times 10^{20}$ joules/y$) = 35.8$ times the average power of the world (see note 5).

236. The NIF laser fusion capsule is approximately 1 millimeter in radius and is filled with cryogenic DT snow (see note 234). It contains $4\pi/3 \times (10^{-3}$ m$)^3 \times 88.8$ kg/m$^3 \times 6.022 \times 10^{23}$ atoms/mole / $(0.001$ kg/mole$) = 2.2 \times 10^{20}$ atoms. Assuming that the capsule burns to completion, the amount of energy released is thus $1/2 \times 2.22 \times 10^{20}$ atoms $\times$ 17.6 MeV/atom $\times 1.602 \times 10^{-13}$ joules/eV $= 3.13 \times 10^8$ joules or 149 sticks of dynamite (see note 113). The radiation dangers only last a week. They mainly come from neutron activation of the apparatus and chamber walls. See S. Sitaraman *et al.*, "Neutron Activation of NIF Final Optics Assemblies," J. Phys. Conf. Ser. **244**, 032025 (2010).

237. See C. M. Braams and P. E. Stott, *Nuclear Fusion: Half a Century of Magnetic Confinement Fusion Research* (Taylor and Francis, 2002); J. D. Lawson, "Some Criteria for a Power Producing Thermonuclear Reactor," Proc. Phys. Soc. B **70**, 6 (1957).

238. A. J. Wootton, H. Y. W. Tsui, and S. Prager, "Edge Turbulence in Tokamaks, Stellarators and Reversed Field Pinches," Plasma Phys. Control. Fusion **34**, 2023 (1992); M. A. Liberman *et al.*, *Physics of High-Density Z-pinch Plasmas* (Springer, 1999); P. R. Garabedian and L.-P. Ku, "Reactors with Stellarator Stability and Tokamak Transport," Fusion Sci. Technol. **47**, 400 (2005); L. C. Woods, *Theory of Tokamak Transport: New Aspects for Nuclear Fusion Reactor Design* (Wiley-VCH, 2006).

239. H. A. Bethe, "The Fusion Hybrid," Physics Today **32**, No. 5, 44 (1979); J. M. Richardson and R. Cohen, eds., *Outlook for the Fusion Hybrid and Tritium-Breeding Fusion Reactors* (National Academies Press, 1987); J. P. Freidberg and A. C. Kadak, "Fusion-Fission Hybrids Revisited," Nature Phys. **5**, 370 (2009).

240. The idea of "leasing" nuclear fuel rods to countries as a way of discouraging their acquisition of nuclear weapons capabilities first appeared in the late 1970s in response to India's first nuclear test. See H. M. Agnew, "Atoms for Lease: An Alternative to Assured Nuclear Proliferation," Bull. Atomic Scientists **32**, No. 5, 22 (1976). The leasing concept resurfaced in 2004 and eventually became the cornerstone of the Bush administration's 2006 Global Energy Nuclear Partnership program. See J.

Deutsch *et al.*, "Making the World Safe for Nuclear Energy," Survival **46**, No. 4, 65 (2004). GNEP had powerful opponents, however, and it was eventually killed. See E. Kintisch, "DOE Asked to Fill in the Blanks on Fuel Recycling Research Plan," Science **312**, 176 (2006); R. L. Garwin, "GNEP: Leap Before Looking," NUCL 61, American Chemical Society Meeting, Chicago, 27 Mar 07; M. B. D. Nikitin, J. M. Parillo and S. A. Squassoni, *Managing the Nuclear Fuel Cycle: Policy Implications of Expanding Global Access to Nuclear Power* (Nova Science, 2008); "Notice of Cancellation of the Global Nuclear Energy Partnership (GNEP) Programmatic Environmental Impact Statement (PEIS)," Federal Resister **74**, 31017 (2009).

241. The countries that presently operate power-oriented reprocessing facilities are France, Britain, Japan, Russia, and India. India is particularly adamant that it has a "right" to reprocess, and South Korea, which does not presently reprocess, has indicated that it wishes to do so. The countries using such facilities on a contract basis include Germany, Switzerland, Holland, Belgium, and Italy. Countries that reprocess only for weapons include the United States, China, and, more recently, North Korea. See K. Ling, "Is the Solution to the U.S. Nuclear Waste Problem in France?," New York Times, 18 May 09; "India Firm on Right to Nuclear Fuel Reprocessing," The Times of India, 20 Dec 09; R. Deshpande, "From Russia with Love: 4 New Nuclear Reactors," The Times of India, 8 Dec 09; J. Vidal, "Thorp Nuclear Plant May Close for Years," The Guardian, 19 May 09; "Seoul Wants 'Sovereignty' in Peaceful Nuclear Development," Chosun Ilbo, 31 Dec 09; "North Korea Produces Plutonium with 'Weapons Potential,'" CNN, 3 Nov 09; N. Nakanishi, "What Happens to Spent Nuclear Fuel?," Reuters, 23 Oct 09; "Time for a New Nuclear Era," JoongAng Ilbo, 5 Jan 10; S.-H. Choe, "U.S. Wary of South Korea's Plan to Reuse Nuclear Fuel," New York Times, 13 Jul 10; "U.S. 'Unlikely to Let S. Korea Reprocess Nuclear Fuel,'" Chosun Ilbo, 14 Jan 10.

242. An early idea along these lines was to distribute rods containing fissile ^{233}U bred from ^{232}Th but spiked with enough ^{238}U so that the ^{233}U couldn't be removed chemically to make weapons (see note 239). See also J. Siegel-Itzkovich, "BGU Combats Nuclear Proliferation," Jerusalem Post, 2 Mar 09.

243. J. Joffe, "Manifest wider die Plutonium-Wirtschaft," Die Zeit, 15 Apr 77; G. Altner and I. Schmitz-Feuerhake, "Die Gefahren der Plutoniumwirtschaft," (Fischer, 1980); F. Barnaby, *Plutonium and Security: The Military Aspects of the Plutonium Economy* (Palgrave Macmillan, 1991); D. Chadwick and A. Steffen, "It's the Plutonium Economy, Stupid," Whole Earth Review, 22 Dec 94; F. B. S. Burnie *et al.*, "Letter: Plutonium Plant Spells More Waste," The Independent, 5 Mar 97; G. Lean, "Decision Times for Britain's Plutonium Economy," The Independent, 28 Jun 98; "Position Paper: Europe Should Lead the Fight Against Global Warming," Greens-EFA Group in the European Parliament, 7 Oct 03.

244. S. Johnson, *The Ghost Map: The Story of London's Most Terrifying Epidemic—and How It Changed Science, Cities and the Modern World* (Riverhead Books, 2006).

245. "Use of Fast Reactors for Actinide Transmutation," International Atomic Energy Agency, IAEA-TECDOC-693, March 1993; V. M. Poplavskii, V. I. Matveev, and N. S. Rabotnov, "Closure of the Nuclear Fuel Cycle: Actinide Balance and Safety," Atomic Energy **81**, 580 (1996); T. Yokoo *et al.*, "Core Performance of Fast Reactors for Actinide Recycling Using Metal, Nitride, and Oxide Fuels," Nucl. Technol. **116**, 173 (1996); T. Inoue, "Actinide Recycling by Pyro-Process with Metal Fuel FBR for Future Nuclear Fuel Cycle System," Prog. Nucl. Energy **40**, 547 (2002).

246. If the world's energy consumption were supplied by nuclear energy, the total mass of fission products produced per year would be 4.4×10^{20} joules/year / (8.82×10^{13} joules/kg) = 4.99×10^6 kg/year (see notes 5 and 219). Assuming that this waste has a specific gravity of about 5 (due to being chemically combined with things like oxygen), the total volume of waste produced in one thousand years is 4.99×10^6 kg/year × 1,000 years / (5,000 kg/m^3) = 1.00×10^6 m^3. An American football field has a length of 109.73 meters including the end zones. The volume of a cubic football field is then $(109.7 \text{ m})^3 = 1.32 \times 10^6$ m^3.

247. Hot nuclear waste was routinely dumped into the world's oceans until the practice was banned by the 1993 London Convention. There remain many sensible arguments for burying waste in undersea sediments, even without first letting it cool off for one thousand years on land. See T. J. Freeman *et al.*, "Modelling Radioactive Waste Disposal by Penetrator Experiments in the Abyssal Atlantic Ocean," Nature **310**, 130 (1984); A. V. Byalko, *Nuclear Waste Disposal: Geophysical Safety* (CRC Press, 1994).

248. Such options have been thought about by a lot of people. See R. Rohmer, *Ultimatum 2* (Dundurn Press, 2004). In this sci-fi thriller, America runs out of places to store its waste and so gives Canada an ultimatum. See also E. E. Angino, G. Dreschhoff, and E. J. Zeller, "Antarctica—A Potential International Burial Area for High-Level Radioactive Wastes," Bull. Eng. Geol. Environ. **13**, 173 (1976). Both nuclear explosions and nuclear waste disposal in Antarctica are presently forbidden by the 1961 Antarctic Treaty. There is actually great concern about radioactive pollution of the high arctic at the moment on account of years of Soviet nuclear dumping, both intentional and unintentional, in waters off the Kola Peninsula and Novaya Zemlya. See W. Sullivan, "Soviet Nuclear Dumps Disclosed," New York Times, 24 Nov 92; B. Salbu, P. Strand and G. C. Christensen, "Dumping of Radioactive Waste in the Barents and Kara Seas," Radiation Protection Dosimetry **62**, 9 (1995); A. V. Yablokov, "Radioactive Waste Disposal in Seas Adjacent to the Territory of the Russian Federation," Marine Pollution Bull. **43**, 8 (2001); J. Vidal, "Russia to Build Floating Arctic Nuclear Stations," The Guardian, 3 May 09.

249. J. O. Jackson, "Soviet Union Gateway to the Gulag," Time, 20 Apr 87; V. Shalamov and J. Glad, *Kolyma Tales* (Penguin, 1995); M. Warren, "'Road of Bones' Where Slaves Perished," Daily Telegraph, 10 Aug 02; M. J. Bollinger, *Stalin's Slave Ships: Kolyma, the Gulag Fleet, and the Role of the West* (Praeger, 2003); A. Apple-

baum, *Gulag: A History* (Anchor, 2004); A. Wolf, "The Big Thaw," Stanford Magazine, 1 Sep 08; A. T. Solzhenitsyn, *Gulag Archipelago* (Harper, 2007).

250. A. Abrahamson, "Mobsters from Ex-Soviet Union a 'Real Threat', Lungren Says," Los Angeles Times, 16 Mar 96; P. Williams, *Russian Organized Crime: The New Threat?* (Routledge, 1997); J. O. Finckenauer and E. J. Waring, *Russian Mafia in America: Culture and Crime* (Northeastern U. Press, 2001); M. Glenny, *McMafia: A Journey through the Global Criminal Underworld* (Knopf, 2008); R. I. Friedman, *Red Mafiya: How the Russian Mob Has Invaded America* (Penguin, 2002); J. Roth, *Die Gangster aus dem Osten* (Europa Verlag, 2003).

CHAPTER 7: CALLING ALL COWS

251. Some of these headlines are from blogs, but most are from print publications. See E. M. Morrison, "Plenty O'Poultry Power," Ag Innovation News, 1 Oct 99; C. Leonard, "Not a Tiger, but Maybe a Chicken in Your Tank," Washington Post, 3 Jan 07; S. Karnowski, "Tons of Turkey Dung to Help Fuel Power Plant in Minnesota," Arizona Daily Star, 24 May 07; "Pig Poo to Help Solve Global Energy Crisis," Times of India, 13 Jun 08; M. Kanellos, "Manure Power Goes Live in Texas," CNET News, 27 Mar 07; E. Gieszl, "Dairy Cow Manure to Power Children's Train Ride," Ultimate Rollercoaster, 14 Aug 08; K. Gashler, "It's Clean, It's Green, It's Cow Manure Power," Elmira Star-Gazette, 11 Aug 09. Other charming headlines on this topic include *Farm's New Juice Isn't Moo, Manure into Gold, Pig Manure Converted to Crude Oil, The Scoop on Poop, Better Fuel through Pig Manure, Tiny Flower Turns Pig Poop into Fuel, The Power of Manure, Government Turns to Manure Power, Power from Cow Poo Heats Homes, Turning Turkey Poop to Power, Bird Dung Is Putting One Small Town on the Map, Turkey Waste Will Power Electric Plant*, and *From Turkey Droppings to Kilowatts*.

252. The following figures are typical of U.S. agriculture estimates: Dairy Cattle 10.0 640.0 6.4, Beef Cattle 7.2 360.0 2.6, Veal 2.3 91.0 0.21, Pigs 8.5 61.0 0.52, Sheep 9.2 27.0 0.25, Goats 11.0 64.0 0.70, Laying Chickens 12.0 1.8 0.022, Broiler Chickens 17.0, 0.9 0.015. The first number is "volatile solids," i.e., the mass of dried manure that will burn, per kilogram of animal, expressed as a multiple of 10^{-3} kilograms per day. (This is typically 20% less than the total mass of dried manure.) The second number is the mass of a typical animal in kilograms. The third is the product of the two in kilograms per head per day. Reasonable guesses for "typical" average values for cattle and chickens of all kinds are 5.0 kg/day and 0.018 kg/day, respectively. See American Society of Agricultural Engineers, "Manure Production and Characteristics," ASAE D384.1, February 2003. The following figures describe world manure production: Cattle 1.36 2.48, Pigs 0.92 0.17, Sheep 1.09 0.10, Goats 0.83 0.21, Chickens 17.90 0.12. The first number is the 2007 animal population, as estimated by the Food and Agriculture Organization of the United Nations (www.fao.org), expressed as a multiple of 10^9. The second number is the mass of

"volatile solids" produced by this population, expressed as a multiple of 10^{12} kilograms per year. The energy content of dried dung is roughly 2.0×10^7 joules/kg, the same as that of wood. See N. R. Kaswan, *Energy Resources and Economic Development: A Study of Rajasthan* (Concept Publishing, 1992). The total energy available in a year's supply of the world's agricultural dung is thus roughly $(2.48 + 0.17 + 0.10 + 0.21 + 0.12) \times 10^{12}$ kg $\times 2.0 \times 10^7$ joules/kg $= 6.2 \times 10^{19}$ joules. This is 48% of the U.S. Energy Information Administration's estimate of 1.3×10^{20} joules that the world obtained from burning coal in 2006 (see note 5).

253. Literature reports of the amount of methane generated by anaerobic digestion of manure vary widely (between 0.19 and 0.47 cubic meter per kilogram of volatile solid) depending on the kind of manure and the digestion conditions. See S. Godbot *et al.*, "Methane Production Potential (B0) of Swine and Cattle Manures—a Canadian Perspective," Environ. Technol. **31**, 1371 (2010); H. B. Moller, S. G. Sommer, and B. K. Ahring, "Methane Productivity of Manure, Straw, and Solid Fractions of Manure," Biomass and Bioenergy **26**, 485 (2004); J. J. H. Huang and J. C. H. Shih, "The Potential of Biological Methane Generation from Chicken Manure," Biotechnol. Bioeng. **23**, 2307 (1981); A. G. Hashimoto, V. H. Varel, and Y. R. Chen, "Ultimate Methane Yield from Beef Cattle Manure: Effect of Temperature, Ration Constituents, Antibiotics, and Manure Age," Agricultural Wastes **3**, 241 (1981). The low end of this range, which is typical of practical digesters, gives $0.2 \text{ m}^3/\text{kg} \times 0.67 \text{ kg/m}^3 \times 5.5 \times 10^7$ joules/kg $/ 2.0 \times 10^7$ joules/kg $= 0.37$ for the fraction of the manure's energy extracted into the methane. If the world's manure were all digested it would thus generate about $0.37 \times 6.2 \times 10^{19}$ joules $= 2.3 \times 10^{19}$ joules per year. This is 19% of the U.S. Energy Information Administration's estimate of 1.1×10^{20} joules that the world obtained from natural gas in 2006 (see note 5).

254. Anecdotal reports indicate that about half the manure mass is typically left over after digestion. This number is consistent with the range of digestion efficiencies (see note 253). The reaction $2 \text{ CHOH} \rightarrow \text{CH}_4 + \text{CO}_2$ gives an average molecular weight for the gas of $(16 + 44)/2 = 30$. At a temperature of 288°K (15°C) the density of this gas is 0.03 kg/mole $\times 1.01 \times 10^5$ pascals $/ (8.314$ joules °K$^{-1} \times 288$°K$) = 1.27$ kg/m^3. The low and high efficiency ranges then give mass fractions of $0.19 \text{ m}^3/\text{kg} \times 1.27 \text{ kg/m}^3 = 0.24$ and 0.60, respectively. Taking the leftover fraction to be 0.5, the power equivalent of the undigested portion of the manure is $0.5 \times 6.2 \times 10^{19}$ joules/y $/ (365$ d/y $\times 24$ h/d $\times 3,600$ sec/h$) = 9.8 \times 10^{11}$ watts, or about 0.53 times the 1.86×10^{12} watts average electric power consumed by the world (see note 190). To make a fair comparison, one should probably reduce this to $0.53 \times 0.36 = 0.19$ to account for electric power generation inefficiencies.

255. The overall energy efficiency of farm animals is difficult to assess because measurements are inaccurate and animals are raised in widely different ways. However, a rough measure may be obtained from the properties of straw, which, when fed to cows, is about 40% indigestible (fiber plus ash). Also, a typical diet for a lactating dairy cow is, after adjusting for water content, about 10 kg. Thus, cow dung

contains roughly half the dry mass, and therefore half the energy, originally fed to the cows. See H. G. Van Pelt, *How to Feed the Dairy Cow* (Fred L. Kimball, 1919).

256. According to the U.S. Department of Agriculture, corn planted for silage in the United States had an average yield in 2007 (a very good year) of 17.5 tons/acre $\times$ 10^3 kg/tonne $\times$ 2.47 acre/ha / (1.1 tons/tonne $\times$ 10^4 m^2/ha) = 3.93 kg/m^3. The total area planted in corn for all purposes was 9.3 $\times$ 10^7 acres $\times$ 10^4 m^2/ha / (2.47 ac/ha) = 3.77 $\times$ 10^{11} m^2. Assuming that corn silage has a dry mass of 35% and that this mass has the energy density of wood, the energy content of all the corn grown in the United States in that year was 0.35 $\times$ 3.93 kg/m^2 $\times$ 3.77 $\times$ 10^{11} m^2 $\times$ 2.0 $\times$ 10^7 joules/kg = 1.04 $\times$ 10^{19} joules. These figures cross-check with the reported yield of 151 bushels per acre and the rule of thumb that corn produces 8 bushels of grain for every ton of silage, per 151 bu/ac / 8 bu/ton = 18.9 tons/ac. Assuming that grain has a dry mass of 85%, the grain constitutes a fraction 0.85 $\times$ 151 bu/ac $\times$ 56 lbs/bu / (0.35 $\times$ 2,000 lbs/ton $\times$ 17.5 tons/ac) = 0.59 of the total dry mass. See "Crop Production 2008 Summary," National Agricultural Statistics Service, U.S. Department of Agriculture, Cr Pr 2-1 (09), January 2009; J. Morrison, E. Nafziger, and L. Paul, "The Relationship between Grain Yield and Silage Yield in Field Corn in Northern Illinois," University of Illinois Extension, 11 Aug 06. According to the U.S. Energy Information Agency's "Energy International Annual," the United States consumed 4.2 $\times$ 10^{19} joules of petroleum energy in 2006. The energy content of the 2007 U.S. corn crop (grain plus stover) is 0.25 times this number.

257. There are many fantastic claims for the biomass productivity of switchgrass, but all properly refereed papers give numbers around 1.0 kg/m^2. Where exactly it lies in the range 0.5–1.2 kg/m^2 depends primarily on rainfall. Taking this number to be dry weight equivalent, the energy yield of switchgrass is about 1.0 kg/m^2 / (0.35 $\times$ 3.90 kg/m^2) = 0.73 times that of corn (see note 256). See K. P. Vogel *et al.*, "Switchgrass Biomass Production in the Midwest USA: Harvest and Nitrogen Management," Agron. J. **94**, 413 (2002); M. R. Schmer *et al.*, "Net Energy of Cellulosic Ethanol from Switchgrass," Proc. Natl. Acad. Sci. **105**, 464 (2008); D. Biello, "Grass Makes Better Ethanol Than Corn Does," Scientific American, 8 Jan 08.

258. The impact of biofuel agriculture on food supplies has already become a political issue (see note 152). See S. Holmes, "Bioenergy: Fuelling the Food Crisis?," BBC News, 4 Jun 08; E. Rosenthal, "U.N. Says Biofuel Subsidies Raise Food Bill and Hunger," New York Times, 7 Oct 08; "Developed Countries' Demand for Biofuels Has Been 'Disastrous,'" The Guardian, 17 Aug 09.

259. Microalgae are single-celled plants. Taxonomists make the following distinctions among them at the class level: *Chlorophyceae* (Green Algae), *Chrysophyceae* (Golden Algae), *Bacillariophyceae* (Diatoms), *Cyanophyceae* (Cyanobacteria). The green algae are eukaryotes with chloroplasts and no pigmentation other than chlorophyll. The golden algae are similar except that they possess, in addition to chlorophyll, other pigments that color them yellow or brown. The diatoms share this complex pigmentation but have, in addition, a silicaceous shell.

The cyanobacteria are called algae for historical reasons but are actually prokaryotes more closely related to bacteria. There are thought to be over 200,000 species of algae. A library of about 125,000 of them is available at www.algaebase.org. The ocean's phytoplankton consists of mostly diatoms. See C. R. Tomas, *Identifying Marine Phytoplankton* (Academic, 1997). The following organisms are commonly mentioned in public domain literature in the context of algal biomass: *Phaeodactylum tricornutum* (Marine Diatom), *Chaetoceros gracilis* (Marine Diatom), *Nitzschia* sp. (Marine Diatoms), *Cyclotella* sp. (Marine Diatoms), *Navicula* sp. (Marine Diatoms), *Amphora* sp. (Marine Diatoms), *Tetraselmis suecica* (Marine Green Alga), *Scenedesmus quadricauda* (Marine Green Alga), *Platymonas* sp. (Marine Green Algae), *Nannochloropsis* sp. (Marine Green Algae), *Botryococcus braunii* (Freshwater Green Alga). The symbol *sp.* means "species" and applies to all species in the specified genus. Not all of the genera listed are purely marine. *Botryococcus braunii*, in particular, has been extensively researched, even though it's a freshwater alga, because of its copious production of lipids suitable for oil industry feedstock. See P. Metzger and C. Largeau, "*Botryococcus braunii*: A Rich Source for Hydrocarbons and Related Ether Lipids," Appl. Microbiol. Biotechnol. **66**, 486 (2005). Most contemporary work on algae for oil production is proprietary.

260. W. Balloni *et al.*, "Mass Cultures of Algae for Energy Farming in Coastal Deserts," in *Energy from Biomass*, A. Straub, P. Chartier, and G. Schleser, eds. (Applied Science, 1983); W. H. Thomas *et al.*, "Yields, Photosynthetic Efficiency and Proximate Composition of Dense Marine Microalgal Cultures," Biomass **5**, 211 (1984); R. A. Anderson, *Algal Culturing Techniques* (Academic Press, 2005); G. Marsh, "Small Wonders: Biomass from Algae," Renewable Energy Focus **9**, 74 (2009); S. Kovalyova, "European Body Sees Algae Fuel Industry in 10–15 Years," Reuters, 3 Jun 09.

261. J. Sheehan *et al.*, "A Look Back at the U.S. Department of Energy's Aquatic Species Program—Biodiesel from Algae," U.S. National Renewable Energy Laboratory, NREL/TP-580-24190, July 1998. This report cites estimates for open-pond production of algal biomass ranging from $40 per tonne to $400 per tonne. (The latter appears on p. 244.) These estimates have many adjustments, including the cost of "feeding" industrial CO_2 to the algae and the government credits for CO_2 capture, so they cannot be compared directly with each other.

262. M. Tampier, "A Sober Look at Biofuels from Algae," Biodiesel Magazine, 9 Mar 09.

263. D. Chaumont, "Biotechnology of Algae Biomass Production: A Review of Systems for Outdoor Mass Culture," J. Appl. Phycology **5**, 593 (1993); E. Molina Grima *et al.*, "Recovery of Microalgal Biomass and Metabolites: Process Options and Economics," Biotechnol. Adv. **20**, 491 (2003); R. M. Knuckey *et al.*, "Production of Microalgal Concentrates by Flocculation and Their Assessment as Aquaculture Feeds," Aquaculture Eng. **35**, 300 (2006).

264. U.S. Office of Technology Assessment, *Energy from Biological Processes: Technical and Environmental Analysis* (Ballinger, 1980); O. Kitani, *Biomass Handbook* (Routledge, 1989).

265. A. L. Mascarelli, "Gold Rush for Algae," Nature **461**, 460 (2009); V. Goel, "Algae Startups Chase Dreams of Fuel from Pond Scum," San Jose Mercury News, 7 Mar 08; J. W. Kram, "PetroSun Commences Algae-Farm Operation," Biodiesel Magazine, 15 Apr 08; A. Jha, "UK Announces World's Largest Algal Biofuel Project," The Guardian, 23 Oct 08; M. LaMonica, "Aurora's Algae Payoff: $50 a Barrel, Plus a Price on Carbon," CNET News, 4 Mar 09; M. Ma, "Redmond Company Beaming Solar Energy to Algae," Seattle Times, 4 May 09; T. Gardner, "PetroAlgae Expects Initial Revenues This Year," Reuters, 14 Aug 09; K. Johnson, "A New Test for Business and Biofuel," New York Times, 16 Aug 09; T. Hsu, "Interest in Algae's Oil Prospects Is Growing," Los Angeles Times, 17 Sep 09; A. C. Mulkern, "Algae as Fuel of the Future Faces Great Expectations—and Obstacles," New York Times, 17 Sep 09.

266. The 2007 Energy Independence and Security Act mandated a renewable fuel standard of 36 billion gallons per year by 2022, 21 billion of which needed to be obtained from cellulosic ethanol and other advanced biofuels. Air Force jets won't run on cellulosic ethanol. The Air Force consumed 3.2 billion gallons of jet fuel in fiscal year 2006. This was 52.5% of all the fossil fuel used by the government that year. See T. Shanker, "Military Plans Tests in Search for an Alternative to Oil-Based Fuel," New York Times, 14 May 06. F. Sissine, "Energy Independence and Security Act of 2007: A Summary of Major Provisions," CRS Report for Congress RL34294, 21 Dec 07.

267. A figure of $33 per gallon widely quoted on blogs allegedly comes from Solix CEO Bryan Wilson.

268. E. Zimmerman, "The Next Green Fuel Source: Algae," CNNMoney, 30 Mar 09.

269. K. Bullis, "Fuel from Algae," Technology Review, 22 Feb 08; P. Gupta "Solazyme Bags U.S. Navy Contract for Green Jet Fuel," Reuters, 24 Sep 09.

270. R. Gold, "Entrepreneurs Wade into the 'Dead Zone,'" Wall Street Journal, 12 Aug 09; R. Gold, "Biofuel Bet Aims to Harvest Fish That Feed on Algae," Wall Street Journal, 18 Aug 09.

271. A. Davis and R. Gold, "U.S. Biofuels Boom Running on Empty," Wall Street Journal, 27 Aug 09.

272. D. Morgan, G. M. Gaul, and S. Cohen, "Farm Program Pays $1.3 Billion to People Who Don't Farm," Washington Post, 2 Jul 06; D. Streitfeld, "As Prices Rise, Farmers Spurn Conservation Program," New York Times, 9 Apr 08.

273. S. Castle, "Europe to Buy 30,000 Tons of Surplus Butter," New York Times, 23 Jan 09.

274. E. Gismatullin and M. Stigset, "Shell, Biopetroleum to Build Algae Plant to Make Fuel," Bloomberg, 11 Dec 07; D. Baker, "Startup Teams with Chevron," San

Francisco Chronicle, 23 Jan 08; B. Sopelsa, "Big Oil Sees Promise in Pond Scum,"
CNBC, 17 Apr 09; A. Jha, "Gene Scientist to Create Algae Biofuel with ExxonMobil,"
The Guardian, 14 Jul 09; K. Howell, "Exxon Sinks $600M into Algae-Based Biofuels
in Major Strategy Shift," New York Times, 14 Jul 09.

275. There has been continual talk since the Carter administration about taxing
"windfall profits" of big oil companies. Various means have been proposed for
doing this, such as eliminating tax breaks and reinstating royalty charges for wells
drilled on public lands. The amounts of extra tax revenue discussed are typically
around $1 billion per year. See S. Lazzari, "The Crude Oil Windfall Profits Tax of
the 1980s: Implications for Current Energy Policy," CRS Report for Congress,
RL33305, 9 Mar 06; S. Hargreaves, "Dems Versus Oil, Part 2," CNN Money, 28 Nov
06; S. Mufson, "Bill Targets Oil Firms and OPEC," Washington Post, 8 May 08.

276. J. Mouawad, "Exxon to Invest Millions to Make Fuel from Algae," New York
Times, 13 Jul 09.

277. An enormous gallery of algae bloom satellite photographs is available at
NASA's Visible Earth website (http://visibleearth.nasa.gov) under "microbiota."

278. Biomass production rates comparable to those of farmed crops are routinely
reported for phytoplankton blooms. See K. R. Arrigo and C. R. McClain, "Spring
Phytoplankton Production in the Western Ross Sea," Science **266**, 261 (1994); M.
Gosselin *et al.*, "New Measurements of Phytoplankton and Ice Algal Production
in the Arctic Ocean," Deep Sea Research, Part II **44**, 1623 (1997). The maximum
rate reported for the Ross Sea was 3.90 gm/m^2-day × 10^{-3} kg/g × 365 days/y = 1.42
kg/m^2 per year. The maximum rate reported for the Chukchi Sea was 0.94 kg/m^2
per year. The biomass production rate for corn is 3.90 kg/m^2 per year (see note
256). This comparison is somewhat misleading because the corn number is a yearly
average, whereas the plankton number is a transient maximum. The maximum
corn production rate is considerably higher than 3.90 kg/m^2 per year, as is the
plankton production rate if the plankton is provided optimal growing conditions
through aquaculture.

279. W. O. Smith and D. M. Nelson, "Phytoplankton Bloom Produced by a Re-
ceding Ice Edge in the Ross Sea: Spatial Coherence with the Density Field," Science
227, 163 (1985); J. J. Stretch *et al.*, "Foraging Behavior of Antarctic Krill *Euphausia
superba* on Sea Ice Microalgae," Mar. Ecol. Prog. Ser. **44**, 131 (1988); M. Spindler,
"Notes on the Biology of Sea Ice in the Arctic and Antarctic," Polar Biology **14**, 319
(1994); M. Gosselin *et al.*, "New Measurements of Phytoplankton and Ice Algal
Production in the Arctic Ocean," Deep Sea Research II **44**, 1623 (1997); G. R. Di-
Tullio *et al.*, "Rapid and Early Export of *Phaeocystis antarctica* Blooms in the Ross
Sea, Antarctica," Nature **404**, 595 (2000); M. P. Lizotte, "Contributions of Sea Ice
to Antarctic Marine Primary Production," American Zoologist **41**, 57 (2001); D.
L. Garrison, K. R. Buck, and G. A. Fryxell, "Algal Assemblages in Antarctic Pack
Ice and in Ice-Edge Plankton," J. Phycology **23**, 564 (1987).

280. D. Malakoff, "Death by Suffocation in the Gulf of Mexico," Science **281**, 190 (1998); D. T. Boesch, "The Gulf of Mexico's Dead Zone," Science **306**, 977 (2004).

281. These algal productivity numbers are often difficult to interpret because algae, like any other agricultural plants, produce differently depending on circumstances, the best of which aren't necessarily cost-effective from the perspective of the grower. One of the higher sustained algae dry biomass production rates reported in the refereed literature is 37 g/m^2-day $\times$ 10^{-3} kg/g $\times$ 365 days/y = 13.5 kg/m^2 per year. Note that this is ash-free dry weight. The plants in question were the common marine green algae *Tetraselmis suecica*. See E. A. Laws *et al.*, "High Algal Production Rates Achieved in a Shallow Outdoor Flume," Biotechnol. Bioeng. **28**, 191 (1986); G. C. Zittelli *et al.*, "Productivity and Photosynthetic Efficiency of Outdoor Cultures of Tetraselmis sussica in Annular Columns," Aquaculture **261**, 932 (2006). According to the U.S. Department of Energy's aquatic species retrospective, field trials for this organism under less controlled circumstances (open pond culture in Roswell, New Mexico) gave summer production rates of 3.8 kg/m^{2}year (10.5 afdw/m^{2}day) but winter rates of zero because the algae died (see note 261). The retrospective concluded that yearly production rates (with other species) of about 3.8 kg/m^{2}year were feasible in large ponds. Corn's dry mass productivity in a good year is 0.35 $\times$ 3.93 kg/m^{2}year = 1.38 kg/m^{2}year (see note 256).

282. According to the U.S. Department of Agriculture, the total area planted in corn for all purposes was 3.77 $\times$ 10^{11} m^2 in 2007 (see note 256). If the energy of this corn is 0.25 times the energy consumed in oil, then algae fields twice as productive as corn fields would require area of 3.77 $\times$ 10^{11} m^2 / (2 $\times$ 0.25) = 7.54 $\times$ 10^{11} m^2 to supply the energy the United States consumed in oil in 2006. The total area of Texas is 6.96 $\times$ 10^{11} m^2.

283. The current market price for corn silage (30% dry weight) is about $20 per ton. This number cross-checks with the current price for shelled corn (85% dry weight) of $3.30 per bushel and the fact that grain constitutes 59% of the silage dry weight: $3.30 bu^{-1} $\times$ 2,000 lbs/ton $\times$ 0.59 $\times$ 0.3 / (56 lbs/bu $\times$ 0.85) = $25 per tonne. The ratio of the coal price to the dry weight silage price is thus about $50 ton^{-1} $\times$ 0.3 / $20 ton^{-1} = 0.75.

284. The chief problem with pure ethanol, or ethyl alcohol (C$_2$H$_5$OH), is that it's miscible with water. This enables us to ruin the fuel (or cheat customers) simply by mixing in some water. Anhydrous ethanol is also hygroscopic, meaning that it will grab water vapor out of the air until it becomes a constant-boiling, or azeotropic, mixture containing 11 molar% (4 weight %) water. Present-day fuels minimize the water problem by mixing ethanol with conventional gasoline. But although water in the tank won't mix with the gasohol, it will draw the ethanol component out of the mixture preferentially, thus forming concentrated ethanol at the bottom of the tank, an effect that can cause serious mischief in an engine.

See A. Johnson, "Mechanics See Ethanol Damaging Small Engines," MSNBC, 1 Aug 08.

285. The fraction of the available energy in sugar siphoned off by the yeast for its own metabolism is difficult to measure and thus not accurately known. The combustion heat of glucose ($C_6H_{12}O_6$) is 2.81×10^6 joules/mole / (0.180 kg/mole) = 1.56×10^7 joules/kg. The combustion heat of ethanol (C_2H_6O) is 1.37×10^6 joules/mole / (0.046 kg/mole) = 2.98×10^7 joules/kg. If the yeast took no energy for itself, the ratio of ethanol mass produced to the glucose mass consumed would be 1.56×10^7 joules/mole / (2.98×10^7 joules/mole) = 0.52. The range for this number usually quoted for actual yeast is between 0.4 and 0.5. Thus, the conversion of glucose to ethanol is thought to be efficient, with the yeast taking a metabolic profit for itself of less than 25%. See S. K. Sharma, K. L. Kaira, and G. S. Kocher, "Fermentation of Enzymatic Hydrolysate of Sunflower Hulls for Ethanol Production and Its Scale-Up," Biomass Bioeng. **27**, 399 (2004).

286. Some newer genetically engineered yeasts will metabolize starch directly, but normally we must convert corn starch to glucose (dextrose) by digesting it with amylase. The amylase may either be purchased commercially or generated by the corn itself through malting.

287. Estimates for the cellulose content of corn stover range from 38 to 70%. Part of the uncertainty is how to count hemicellulose, a collection of heteropolymers similar to cellulose that vary from plant to plant. Ordinary wood is about 40% cellulose by weight. See R. M. Rowell, *Handbook of Wood Chemistry and Wood Composites* (CRC Press, 2005).

288. Microalgae have been studied as potential substitutes for wood chips in paper manufacture. Their cellulose contents are low, typically about 10% by mass. Hemicelluloses make up another 10%. See C. Ververis *et al.*, "Celluloses, Hemicelluloses and Ash Content of Some Organic Materials and Their Suitability for Use as Paper Pulp Supplements," Bioresource Technol. **98**, 296 (2007).

289. The first trees are usually dated as late Devonian, about 380 million years ago. See L. Smith, "Fossil from a Forest That Gave Earth Its Breath of Fresh Air," London Times, 19 Apr 07.

290. Hydrolysis of cellulose to glucose is slightly exothermic, the proof of this being that cellulase enzymes accomplish their task without being supplied energy externally. See N. Karim and S. Kidikoro, "Precise and Continuous Observation of Cellulase-Catalyzed Hydrolysis of Cello-Oligosaccharides Using Isothermal Titration Calorimetry," Thermochim. Acta **412**, 91 (2004).

291. Q. Xu *et al.*, "Cellulases for Biomass Conversion" and A. Miettinen-Oinen, "Cellulases in the Textile Industry," in *Industrial Enzymes: Structure, Function and Application*, J. Polaina and A. P. MacCabe, eds. (Springer, 2007); V. A. Nierstrasz and M. M. C. G. Warmoeskerken, "Process Engineering and Industrial Enzyme Application," in *Textile Processing with Enzymes*, A. Cavaco-Paulo and G. M. Gübitz, eds. (CRC Press, 2003); S. Singh, R. C. Kuhad, and O. P. Ward, "Industrial Ap-

plications of Microbial Cellulases," in *Lignocellulose Biotechnology: Future Prospects,* R.C. Kuhad and A. Singh, eds. (I.K. International, 2007). The $1.1 billion estimate for the world industrial enzyme market refers to the "technical" industrial enzyme market. Food and agriculture enzymes, which make up an additional $1.2 billion, are sometimes also referred to as industrial. See "BCC Research Report Indicates Global Market for Industrial Enzymes Worth $2.7 Billion," Reuters, 5 Dec 07. Several sources quote the cellulase market at around $140 million per year.

292. F. Schauer and R. Borriss, "Biocatalysis and Biotransformation," in *Advances in Fungal Biotechnology for Industry, Agriculture and Medicine,* J. S. Tkacz and L. Lange, eds. (Kluwer Academics/Plenum, 2004); L. Xia and X. Shen, "High-Yield Cellulase Production by *Trichoderma reesei* ZU-02 on Corn Cob Residue," Bioresource Technol. **91**, 259 (2004).

293. The true enzyme costs for cellulosic ethanol manufacture are difficult to assess because of the political pressures to understate them. Xu *et al* (see note 291). estimated a cost of $5 per gallon of ethanol in 2007 but then quoted a conference proceedings claim by enzyme manufacturer Genencor of a breakthrough lowering this cost to $0.20 per gallon. Because the enzyme in question is the same as that used in textile manufacture, such a breakthrough should have caused a catastrophic price collapse of the textile enzyme market—which it did not, in fact, do. Genencor has an ongoing grant from the U.S. government to develop low-cost enzymes for cellulosic ethanol manufacture. An even lower cost of $0.10 is quoted in A. Aden *et al.*, "Lignocellulosic Biomass to Ethanol Process Design and Economics Utilizing Co-Current Dilute Acid Prehydrolysis and Enzymatic Hydrolysis for Corn Stover," National Renewable Energy Laboratory, NREL/TP-510-32438, June 2002.

294. The reason they're hard to find is that the assays in question are difficult and expensive, not because the chemistry is subtle. Assuming that corn is 60% starch by weight, the current market price of starch (glucose) energy is (see note 285) $3.30 bu^{-1} × 2.2 lbs/kg / (65 lbs/bu × 0.60 × 1.56 × 10^7 joules/kg) = $1.19 × 10^{-8} joule^{-1}. The cost of ethanol energy is $1.64 gal^{-1} × 264 gals/m^3 / (798 kg/m^3 × 2.97 × 10^7 joules/kg) = $1.83 × 10^{-8} joule^{-1}. Thus, a measurement accurate only to 10% could resolve the matter.

295. Alternates to cellulosic digestion include acid hydrolysis and conventional syngas manufacture. See P. C. Badger, "Ethanol from Cellulose: A General Review," in *Trends in New Crops and New Uses,* J. Janick and A. Whipkey, eds. (ASHS Press, 2002); H. G. Lawford and J. D. Rousseau, "Cellulosic Fuel Ethanol," Appl. Biochem. Biotechnol. **106**, 457 (2003); J. Chu, "Reinventing Cellulosic Ethanol Production," Technology Review, 10 Jun 09; S. Ashley, "Newly Uncovered Enzymes Turn Corn Plant Waste into Biofuel," Scientific American, 22 Jul 09.

296. This is from Mark Twain's short story "The Private History of a Campaign That Failed," published in many anthologies.

297. Brewer's yeast *Saccharomyces cerevisiae* will tolerate about a 0.2 volume fraction of ethanol. Water and ethanol mix with insignificant volume change, so this

corresponds to a mass fraction of 0.2×798 kg/m^3 / $(0.8 \times 1000$ kg/m^3 + 0.2×798 kg/m^3) = 0.17. See F.-M. Lee and R. H. Pahl, "Solvent Screening Study and Conceptual Extractive Distillation Process to Produce Anhydrous Ethanol from Fermentation Broth," Ind. Eng. Chem. Process Des. Dev. **24**, 168 (1985); T. Paalme *et al.*, "Growth Efficiency of *Saccharomyces cerevisiae* on Glucose/Ethanol Media with a Smooth Change in the Dilution Rate," Enzyme Microb. Technol. **20**, 174 (1997).

298. Distillation is a notoriously inefficient technology. The theoretical minimum energy required to concentrate ethanol is constrained by its entropy of mixing S = $-v$ R (f ln (f) + (1 − f) ln (1-f)), where v is the number of moles (of both types), R is the ideal gas constant, and f is the initial fraction of the molecules that are ethanol. For the case of 20% ethanol by volume (17% by weight), f = 0.17 / (0.83 × 46/18 + 1) = 0.054 and S = 0.21 v R (see note 297). The minimum energy cost per kilogram of ethanol separated at room temperature (300°K) is 0.21 × 8.31 joules mole^{-1}°K^{-1} × 300°K / (0.054 × 0.048 kg/mole) = 2.02 × 10^5 joules/kg. We could achieve this ideal if we had a perfect reverse osmosis technology, but we don't. Separating ethanol by distillation, by contrast, requires expending the heat of vaporization of ethanol (3.86 × 10^4 joules/mole) and water (4.07 × 10^4 joules/mole) in some proportions multiple times, for each distillation is imperfect. Reaching the azeotropic concentration (96% alcohol by volume) requires roughly ten distillations. In the first distillation, the molar ethanol fraction increases from 0.054 to 0.35. The heat required to vaporize this mixture, per kilogram of ethanol vaporized, is (0.35 × 3.86 + 0.65 × 4.07) × 10^4 joules/mole / (0.35 × 0.046 kg/mole) = 2.48 × 10^6 joules/kg. The physical reason the amount of heat is so much greater than the ideal value is that the temperature differences between the vaporization and condensation steps is small. For example, in the first distillation the temperature difference is about 8°K, so the amount of entropy removed by a given amount of energy is reduced from the ideal value by the factor 8°K/352°K = 0.022. Energy is also required to remove the 4% water left at the azeotrope, for which distillation can't be used.

299. D. Pimentel, "Ethanol Fuels: Energy Balance, Economics and Environmental Impacts Are Negative," Natural Resources Research **12**, 127 (2003). This paper assumes an ethanol energy content of 7.7 × 10^4 btu/gal × 1055 joules/btu × 264 gals/m^3 / 798 kg/m^3 = 2.69 × 10^7 joules/kg, slightly lower than ethanol's true combustion heat of 2.98 × 10^7 joules/kg (see note 285). It estimates that the coal burned to produce a gallon of ethanol releases 39,067 btu of heat (see note 298). The ratio of this consumed energy to the energy contained in the gallon of ethanol it helps produce is 3.91 × 10^4 btu/gal / (7.7 × 10^4 btu/gal) = 0.51.

300. Most living things have mastered it on a small scale, however. Mitochondria, the organelles in eurcaryotes that manufacture ATP from other energy sources such as sugars, use an electrochemical potential (a voltage) across their membranes as an intermediate step in the process. The ATP synthase molecule is, in effect, a tiny electric motor embedded in the mitochondrion's membrane. The potential drop across the membrane is about 0.18 volt. See L. B. Chen, "Mitochondrial Membrane Potential in Living Cells," Annu. Rev. Cell Biol. **4**, 155 (1988).

301. The generation of synthesis gas from biomass (or coal) is an endothermic process that requires first burning fuel to raise the temperature. This burning step stores the needed reaction energy as heat. One can, however, skip this burning by processing the biomass through baths of molten metal kept hot electrically. See R. Gavin, "Betting on a Hot Market for Syngas," Boston Globe, 25 Aug 08.

302. D. W. Thayer, "Facultative Wood-Digesting Bacteria from the Hind-Gut of the Termite *Reticulitermes hesperus*," J. Gen. Microbiol. **95**, 287 (1976); R. Radek, "Flagellates, Bacteria and Fungi Associated with Termites: Diversity and Function in Nutrition—a Review," Ecotropica **5**, 183 (1999); Y. Hongoh *et al.*, "Genome of an Endosymbiont Coupling N2 Fixation to Cellulolysis within Protist Cells in Termite Gut," Science **332**, 1108 (2008); M. E. Scharf and A. Tartar, "Termite Digestomes as Sources for Novel Lignocellulases," Biofuels Bioproducts and Biorefining **2**, 540 (2008); C. Y. Johnson, "Termites Show Complexity of Biofuel Work," Boston Globe, 8 Dec 08.

303. N. Savage, "Making Gasoline from Bacteria," Technology Review, 1 Aug 07; C. Ayres, "Scientists Find Bugs That Eat Waste and Excrete Petrol," London Times, 14 Jun 08.

304. Some microbes eat petroleum. See W. H. Thorpe, "Petroleum Bacteria and the Nutrition of *Psilopa petrolei*," Nature **130**, 437 (1932); R. W. Stone, M. R. Fenske, and A. G. C. White, "Bacteria Attacking Petroleum and Oil Fractions," J. Bacteriol. **44**, 169 (1942); S. Hino, K. Watanabe, and N. Takahashi, "Isolation and Characterization of Slime-Producing Bacteria Capable of Utilizing Petroleum Hydrocarbons as Sole Carbon Source," J. Fermentation and Bioeng. **84**, 528 (1997); S. Roy *et al.*, "Survey of Petroleum-Degrading Bacteria in Coastal Waters of Sunderban Biosphere Reserve," World J. Microbiol. Biotechnol. **18**, 575 (2002); J.-R. Chong, "Tar Pits' Secret Bubbles Up," Los Angeles Times, 14 May 07; S. B. Akinde and O. Obire, "Aerobic Heterotrophic Bacteria and Petroleum-Utilizing Bacteria from Cow Dung and Poultry Manure," World J. Microbiol. Biotechnol. **24**, 1999 (2008); M. Teramaoto *et al.*, "*Oceanobacter*-Related Bacteria Are Important for the Degradation of Petroleum Aliphatic Hydrocarbons in the Tropical Marine Environment," Microbiol. **155**, 3361 (2009).

CHAPTER 8: TRASH ASH

305. Archaeological knowledge of ancient trash practices extends back to neolithic kitchen middens. A case has been made that municipal dumps as we know them were invented at the dawn of civilization as a necessity for dealing with sedentary life. See T. Hardy-Smith and P. C. Edwards, "The Garbage Crisis in Prehistory: Artefact Discard Patterns at the Early Natufian Site of Wadi Hammeh 27 and the Origins of Household Refuse Disposal Strategies," J. Anthropol. Archaeol. **23**, 235 (2004).

306. According to the U.S. Environmental Protection Agency, the United States generated 2.51×10^8 tons $\times$ 909 kg/ton = 2.28×10^{11} kg of municipal solid waste in

2006. See "Municipal Solid Waste in the United States: 2006 Facts and Figures," U.S. Environmental Protection Agency, Office of Solid Waste, EPA-530-F-07-030, November 2007. Here is a rough breakdown of the EPA estimates: Paper 0.339 0.94 0.44, Glass 0.053 1.00 0.00, Metals 0.076 1.00 0.00, Plastic 0.117 1.00 0.85, Rubber and Leather 0.026 0.90 0.85, Textiles 0.047 0.90 0.44, Wood 0.055 0.80 0.44, Other 0.018 0.50 0.44, Food Waste 0.125 0.30 0.44, Yard Trimmings 0.129 0.40 0.44, Miscellaneous 0.015 1.00 0.44. The first number is the component's mass fraction in the raw waste. The second number is the fraction of the component's mass that isn't water. The third number is the fraction of the component's dried mass that is carbon. Most of dried mass consists of cellulose $(C_6H_{10}O_5)_n$, which has a carbon mass fraction of $6 \times 12 / (6 \times 12 + 10 + 5 \times 16) = 0.44$. The other major component is plastic $(CH_2)_n$, which has a carbon fraction of $12/14 = 0.85$. Multiplying the first two numbers and adding gives 0.79, the mass fraction of the raw waste that isn't water. Multiplying all three numbers and adding gives 0.35, the mass fraction of the raw waste that is carbon. The quotient $0.35/0.79 = 0.44$, the carbon mass fraction of dried waste, cross-checks against the sequestration fraction of 0.25 reported in B. F. Staley and M. A. Barlaz, "Composition of Municipal Solid Waste in the United States and Implications for Carbon Sequestration and Methane Yield," J. Environ. Eng. **135**, 901 (2009). Their number is smaller because they subtracted the carbon mass lost from landfills due to anaerobic methane generation, per 2 $C H_2O \rightarrow CH_4 + CO_2$. Thus, according to the EPA report, the United States generated $0.35 \times 2.28 \times 10^{11}$ kg $= 7.98 \times 10^{10}$ kg of trash carbon in 2006, or 88 million tons. According to the U.S. Energy Information Agency's "International Energy Annual," the United States consumed in 2006 1.1×10^9 tons $\times$ 909 kg/ton $= 1.00 \times 10^{12}$ kg of carbon in the form of coal, $12/14 \times 2.07 \times 10^7$ bbl/day $\times$ 365 days $\times$ 139 kg/bbl $= 9.00 \times 10^{11}$ kg of carbon in the form of oil, and $12/16 \times 2.17 \times 10^{13}$ ft^3 $\times$ 0.67 kg/m^3 / (3.28 ft/m)^3 $= 3.09 \times 10^{11}$ kg of carbon in the form of natural gas. The mass of trash carbon the United States generated in 2006 was 0.080, 0.089, and 0.258 times these numbers, respectively. See also R. Malone, "World's Worst Waste," Forbes, 24 May 06.

307. Taking the U.S. trash generation and energy consumption to be typical of the entire world when it industrializes, the amount of carbon in trash will be about $(1/0.080 + 1/0.089 + 1/0.258)^{-1} = 0.036$ times the amount of carbon released into the air as carbon dioxide through fossil fuel burning (see note 306). Assuming that this burning continues for another two hundred years, the time it would take to return the carbon dioxide thus released to the ground in the form of unburned buried trash would be 200 years/0.036 = 5,556 years.

308. The many bullish reports in the press about extracting energy from trash, either as electricity generated through incineration or as methane given off by landfills, usually fail to convey how small this energy is compared to all fossil fuel consumption. See D. Chen, "Converting Trash Gas into Energy Gold," CNN, 5 May 06; T. K. Grose/Newport, "Britain to Burn Trash for Energy," Time, 9 Jun 08; J.

Rather, "Tapping Power from Trash," New York Times, 13 Sep 08; P. McKenna, "Could Trash Solve the Energy Crisis?," ABC News, 26 Apr 09; J. Ball, "Garbage Gets a Fresh Look as Source of Energy," Wall Street Journal, 15 May 09.

309. The components of trash that are cellulosic (dry carbon fraction 0.44) have an energy content of about 2.0×10^7 joules per kilogram of dry mass (see note 306). The components that are fatty (dry carbon fraction of 0.85) have an energy content of roughly 4.4×10^7 joules per kilogram of dry mass. The energy content of trash generally is thus about 5.0×10^7 joules per kilogram of carbon. The U.S. Environmental Protection Agency's estimates then give 7.98×10^{10} kg $\times 5.0 \times 10^7$ joules/kg $= 3.99 \times 10^{18}$ joules for the energy content of the trash that the United States generated in 2006 (see note 306). According to the U.S. Energy Information Agency's "International Energy Annual," the United States consumed 2.25×10^{16} btu $\times 1,055$ joules/btu $= 2.37 \times 10^{19}$ joules of coal energy and $(2.25 + 4.00 + 2.22) \times 10^{16}$ btu $\times 1,055$ joules/btu $= 8.94 \times 10^{19}$ joules of total fossil fuel energy in that year. The energy contained in the trash generated by the United States in 2006 was 0.17 times the first of these numbers and 0.04 times the second.

310. Present-day trash incinerators that make electricity still have difficulty competing with landfill disposal even after their capital costs are discharged. See B. Barber, "Tulsa Burn Plant to Shut Down Monday," Tulsa World, 21 Jul 07. This particular plant was restarted a year later. See I. Brat, "Cities Give Waste-to-Energy Plants a Second Look," Wall Street Journal, 6 Dec 08. Also, environmental opposition continues to block new waste-to-energy facilities, which it would have trouble doing if the facilities were highly profitable. See E. Gies, "Waste-to-Energy Plants a Waste of Energy, Recycling Advocates Say," New York Times, 4 Jul 08; T. Hardy and C. Bowman, "Sacramento Trash-to-Energy Plan Raises Red Flags," Sacramento Bee, 17 Nov 09; C. S. Psomopolous, A. Bourka, and N. J. Themelis, "Waste-to-Energy: A Review of the Status and Benefits in USA," Waste Management **29**, 1718 (2009).

311. The (dry) mass of wood consumed in the United States in 1999 was 2.3×10^8 tonnes. See J. L. Howard, "U.S. Timber Production, Trade, Consumption and Price Statistics 1965–1999," U.S. Forest Service, Forest Products Laboratory, FPL-RP-595, April 2001. According to the U.S. Energy Information Agency's "International Energy Annual," the United States consumed 1.038×10^9 tons / (1.1 tons/tonne) $= 9.44 \times 10^8$ tonnes of coal that year. The energy content per unit mass of the two fuels is roughly the same, so the energy that would have been obtained by burning all the wood consumed in the United States in 1999 is 2.3×10^8 tonnes / (9.44×10^8 tonnes) $= 0.24$ times the energy the United States obtained that year from coal.

312. R. Gold, "Wood Pellets Catch Fire as Renewable Energy Source," Wall Street Journal, 7 Jul 09; C. Mortished, "Shell Turns to Wood Chips and Straw in Search for the Road Fuel of the Future," London Times, 2 Mar 07.

313. The annual production of artificial polymeric materials can be gauged roughly by the amount of ethane/ethylene sold. Here is a breakdown of refined petroleum

product consumption by the United States in 2008 from the U.S. Energy Information's Petroleum information website (in units of 1,000 bbl): Gas Liquids 748,246, Other Liquids −9,203, and Finished Petroleum Products 6,397,212. These add to Total Petroleum Products 7,136,255. The relevant subcategories of Gas Liquids were Ethane/Ethylene 250,783 and Propane/Propylene 422,439. The relevant subcategories of Finished Petroleum Products were Petrochemical Feedstocks 202,129, Finished Motor Gasoline 3,290,057, Distillate Fuel Oil 1,444,024, and Kerosene-Type Jet Fuel 563,111. Ethane is easily converted to ethylene by steam cracking, so the two are effectively the same thing. Ethane, unlike propane, isn't consumed as fuel, so it is a good measure of petroleum input to manufacture of polyethylene and polyethylene terephthalate (PET), two of the three main components of municipal trash plastic, the other being polypropylene. The meaning of one bbl of ethane/ethylene is unfortunately ambiguous, because both substances are gases at room temperature. Assuming it to mean the equivalent volume of cryogenic liquid (specific gravity 0.57), the total mass of ethane/ethylene produced in the United States in 2008 was 2.51×10^8 bbl $\times 570$ kg/m^3 / (6.29 bbl/m^3) = 2.27×10^{10} kg. This number is consistent with the American Chemical Council (www.americanchemistry.com) estimates for total 2008 North American sales of polyethylene (low-density, linear low-density, and high-density) and thermoplastic polyester (PET) of $(7.14 + 12.39 + 16.82 + 9.81) \times 10^9$ lbs / (2.2 lbs/kg) = 2.1×10^{10} kg. Their estimate for 2008 sales of resins of all types was 1.04×10^{11} lbs / (2.2 kg/lb) = 4.73×10^{10} kg. See also *The Resin Review* (American Chemistry Council, 2009). This cross-checks with the U.S. Environmental Protection Agency's estimate of the total mass of plastic in U.S. municipal solid waste in 2006: $0.117 \times 2.28 \times 10^{11}$ kg = 2.67×10^{10} kg (see note 306). The American Chemical Council's 2008 total North American resin sales number was 4.73×10^{10} kg / (1.08×10^{12} kg) = 0.044 times the BP Statistical Review of World Energy's estimate for the mass of oil consumed by the United States in 2008. The fraction of oil's energy diverted into plastic was 0.044 as well, because the weighted average of the combustion heats of the major components of trash plastic is roughly the same as that of crude oil. Here are the important combustion heats expressed in multiples of 10^7 joules/kg: Polyethylene 4.46, Polypropylene 4.27, Polystyrene, 4.20, PET 2.23. See R. N. Walters, S. M. Hackett, and R. E. Lyon, "Heats of Combustion of High Temperature Polymers," *Fire and Materials* **24**, 245 (2000); M. Llanos, "Plastic Bottles Pile Up as Mountains of Waste," MSNBC, 3 Mar 05.

314. "Florida County Plans to Vaporize Landfill Trash," USA Today, 9 Sep 06; M. Behar, "The Prophet of Garbage," Popular Science, 1 Mar 07; T. McNicol, "The Ultimate Garbage Disposal," Discover, 18 May 07; S. Malone, "Modern-Day Alchemy: Turning Trash into Power," Reuters, 20 May 08; P. Fairley, "Garbage In, Megawatts Out," Technology Review, 2 Jul 08; E. Bland, "Plans to Burn Trash with Plasma Face Hurdles," MSNBC, 15 Dec 08; K. Bullis, "Converting Garbage into Fuel," Technology Review, 27 May 09.

315. The present retail cost of motor gasoline energy is roughly \$3.50 gal^{-1} × 264 gal/m^3 / (730 kg/m^3 × 4.4 × 10^7 joules/kg) = \$2.88 × 10^{-8} per joule. The present retail cost of electric energy is roughly \$0.10 kwh^{-1} / (10^3 wh/kwh × 3,600 sec/h) = \$2.78 × 10^{-8} per joule.

316. Coal burning is a greater environmental problem than trash burning, mainly because the amount of coal is greater than the amount of trash. Several sources quote the amount of coal ash generated annually in the United States as about 1.3 × 10^8 tons × 909 kg/ton = 1.18 × 10^{11} kg (see note 306). See S. Dewan, "Hundreds of Coal Ash Dumps in the U.S. Lack Regulation," New York Times, 7 Jan 09. This number is consistent with estimates of 10% by weight for the typical ash content of coal and the U.S. Energy Information Administration's "International Energy Annual" estimate for 2006 U.S. coal consumption of 0.1 × 1.1 × 10^9 tons × 909 kg/ton = 1.00 × 10^{11} kg. See E. E. Somemeier, *Coal: Its Composition, Analysis, Utilization and Valuation* (BiblioBazaar, 2008). Even if recyclable metal and glass had not been removed from the trash, so that 20% of its mass were noncombustible, the amount of ash generated by burning all U.S. trash in 2006 would still have been only 0.2 × 2.28 × 10^{11} kg = 4.56 × 10^{10} kg. According to the U.S. Environmental Protection Agency, the mass fraction of U.S. trash incinerated for all reasons in 2006 was some number greater than 0.13, the fraction incinerated for energy recovery (see note 306). *Waste Incineration and Public Health* (National Academies Press, 2000) estimated this fraction to be 0.16 in 1993. Even with a generous increase of this number to 0.2, the trash ash mass produced by the United States in 2006 would be only 0.2 × 4.56 × 10^{10} kg / 1.18 × 10^{11} kg = 0.08 times the coal ash mass it produced that year. The metal mass fraction of coal ash varies widely depending on the source, but is typically 2.0 × 10^{-4} (200 ppm) for common impurities such as Zn, Ni, Cu, Co, Cr, Mn, Sr, V, and Pb. Only the last of these is severely poisonous. The amount of Fe is very large, typically a mass fraction of 0.12. See, for example, K. A. Katrinak and S. A. Benson, "Trace Metal Content of Coal and Ash as Determined Using Scanning Electron Microscopy with Wavelength-Dispersive Spectrometry," Preprints of Papers, American Chemical Society, Division of Fuel Chemistry **40**, 798 (1995); C. A. Palmer, ed., "The Chemical Analysis of Argonne Premium Coal Samples," U.S. Geological Survey Bulletin 2144, 1997; *Managing Coal Combustion Residues in Mines* (National Academies Press, 2006). The metal content of trash is generally comparable to that of coal, although it can be several times larger, especially for very dangerous metals such as Pb and Cd. See D. Van Buren, G. Poe, and C. Castaldini, "Characterization of Hazardous Waste Incinerator Residuals," U.S. Environmental Protection Agency, Hazardous Waste Engineering Research Laboratory, EPA/600/52-87/017, May 1987. Hg doesn't show up in ash because it volatilizes in the combustion process. See L. V. Heebink and D. J. Hassett, "Release of Mercury Vapor from Coal Combustion Ash," J. Air Waste Manage. **52**, 927 (2002). Coal also contains immense amounts of sulfur, which trash does not. There's a scarcity of precise measurements of coal's sulfur content in the public-domain

literature, but a rough estimate for the average is the U.S. Department of Energy's boundary between "high-sulfur" and "medium-sulfur" coals: 1.67×10^{-6} lbs/btu $\times$ 2.0×10^7 joules/kg / (2.2 lbs/kg $\times$ 1055 joules/btu) = 0.014 mass fraction. See "U.S. Coal Reserves: An Update by Heat and Sulfur Content," U.S. Energy Information Administration, DOE/EIA-0529(92), February 1993; J. T. Dunham, C. Rampacek, and T. A. Henrie, "High-Sulfur Coal for Generating Electricity," Science **184**, 346 (1974). The sulfur dioxide produced when coal burns, if not scrubbed out, rains out as sulfuric acid. Had all of the sulfur in the coal the United States burned in 2006 been emitted into the air, it would have generated $(32 + 4 \times 16 + 2)/32 \times 0.014 \times 1.0$ $\times 10^{12}$ kg = 4.29×10^{10} kg of H_2SO_4. To put this number in perspective, it's about half the total mass of coal ash generated in 2006, or ten times the amount of ash that would have been generated by burning all U.S. trash that year. According to the U.S. Environmental Protection Agency, the United States emitted 1.02×10^7 tons $\times$ 909 kg/ton = 9.27×10^9 kg of H_2SO_4 in 2005. However, this conclusion is based on numbers submitted by participating companies themselves as part of the sulfur dioxide cap-and-trade process, so it is biased in complex ways. See "Acid Rain Program: 2005 Progress Report," U.S. Environmental Protection Agency, Office of Air and Radiation, EPA-430-R-06-015, October 2006; Z. Coile, "Cap-and-Trade Model Eyed for Cutting Greenhouse Gases," San Francisco Chronicle, 3 Dec 07. Some of the reported emission decline came from scrubbing, which involves rendering the SO_2 in the exhaust gas to CaSO4 (gypsum) using lime. But scrubbing equipment is expensive, and some companies chose instead to burn low-sulfur coal, in some cases from foreign sources. See M. Clayton, "Why Coal-Rich US Is Seeing Record Imports," Christian Science Monitor, 10 Jul 06. The severe public relations problems of coal sometimes have amusing consequences. See M. Hvistendahl, "Coal Ash Is More Radioactive Than Nuclear Waste," Scientific American, 13 Dec 07; M. Cimitile, "Is Coal Ash in Soil a Good Idea?," Scientific American, 6 Feb 09; S. Goldenberg, "Obama Administration Criticised Over Failure to Disclose Coal Dump Locations," The Guardian, 18 Jun 09; S. Dewan, "E.P.A. Lists 'High Hazard' Coal Ash Dumps," New York Times, 30 Jun 09.

317. New York City's practices are a notorious case in point. New York abandoned incineration of garbage in 1999, partly on account of the difficulties of disposing of the ash, but mostly because of steep capital and operating costs. See D. Martin, "City's Last Waste Incinerator Is Torn Down," New York Times, 6 May 99. A major part of the cost problem was legal, however, notably the 1994 Supreme Court decision declaring incinerator ash to be toxic waste. New York City subsequently abandoned what was then the largest trash disposal site in the world, the Fresh Kills landfill, to settle a lawsuit brought by the Borough of Staten Island under the Federal Clean Air Act. Fresh Kills is now a park. New York did not, however, change its method of disposing of trash but simply arranged to ship it to landfills out of state, notably to Virginia, at an additional cost to itself of $300 million per year. See V. S. Toy, "Staten Island Leadership Sues to Close Fresh Kills,"

New York Times, 26 Mar 96; E. Lipton, "As Garbage Piles Up, So Do Worries," Washington Post, 12 Nov 98; K. Johnson, "Dumping Ends at Fresh Kills, Symbol of a Throw-Away Era," New York Times, 18 Mar 01; I. Urbina, "City Trash Plan Forgoes Trucks, Favoring Barges," New York Times, 7 Oct 04; J. Rather, "A Long, Long Haul from the Curb," New York Times, 4 Dec 05; S. Chan, "A New Scenic Destination: That's Right, It's Fresh Kills," New York Times, 8 Apr 06. Smaller versions of this drama occurred all over the United States. See S. Bronstein, "Garbage Landfills in Half of the States Approaching Capacity," New York Times, 12 Feb 87; I. Peterson, "Mounds of Garbage Signal Landfill Crisis in Jersey," New York Times, 16 Apr 87; B. Yap, "Hawaii's Next Big Export: Municipal Trash," USA Today, 14 Jul 08.

318. Cost constraints show up dramatically in recycling programs, which are popular with voters but more expensive than simple disposal of the recyclable items in landfills. Exactly how much more is difficult to quantify because of hidden subsidies, but estimates of the collection cost for plastic have run as high as fifteen times the market value of the plastic collected. New York City temporarily suspended the recycling of glass and plastic in 2002 for cost reasons. See J. Holusha, "Who Foots the Bill for Recycling?," New York Times, 25 Apr 93; D. Cardwell, "New Yorkers Getting Used to Recycling a Little Less," New York Times, 12 Aug 02; M. Cooper, "City to Resume Recycling of Plastics," New York Times, 15 Jan 03.

319. Extremely sloppy incineration is mercifully rare. See A. Levine, "Effects of Toxic Smoke Worry Troops Returning from Iraq," CNN, 15 Dec 08; K. Bradsher, "China's Incinerators Loom as Global Hazard," New York Times, 11 Aug 09.

320. Chlorine is a critical component of dioxins and furans, which are particularly poisonous products of incomplete burning related to polycyclic aromatic hydrocarbons (see note 114). The most infamous of them is 2,3,7,8-tetrachlorodibenzo-p-dioxin, or "dioxin" for short. Its production in the burning of chlorinated plastics, notably in backyard barrels and municipal trash incinerators, is an important worldwide environmental problem. See A. Schecter *et al.*, "Dioxins: An Overview," Environ. Res. **101**, 419 (2006); M. R. Beychok, "A Data Base for Dioxin and Furan Emissions from Municipal Refuse Incinerators," Atmos. Environ. **21** 29 (1987). The metal menace in smoke comes mainly through fly ash. High-end incinerators minimize fly ash in exhaust by precipitating it out electrostatically and adding it to bottom ash for disposal. See W. R. Niessen, *Combustion and Incineration Processes* (CRC Press, 2002); R. Zevenhoven and P. Kilpinen, *Control of Pollutants in Flue Gases and Fuel Gases*, Laboratory of Energy and Environmental Protection, Helsinki University of Technology, TKK-ENY-4, February 2003.

321. B. Harden, "Japan's Trash Technology Helps Deodorize Dumps in Tokyo," Washington Post, 18 Nov 08.

322. Transportation fuel costs have particularly high leverage in cases like New York City's, in which trash is transported to landfill sites in other states. Collection costs for trash are generally higher than landfill costs, although most of the collection

cost is salaries. See F. Kreith and G. Tchobanoglous, *Handbook of Solid Waste Management* (McGraw-Hill, 2002). Anecdotal evidence from small-town newspapers is that fuel costs for hauling are between 15 and 20% of the total disposal cost. See, for example, J. Bencks, "Waste Haulers Raise Fuel Surcharge as Prices Rise," North Andover Eagle-Tribune, 7 Aug 08.

323. T. Lieuwen, V. Yang, and R. Yetter, eds., *Synthesis Gas Combustion: Fundamentals and Applications* (CRC Press, 2009).

324. The public relations difficulties of incinerator ash are nicely illustrated by L. Greenhouse, "Justices Decide Incinerator Ash Is Toxic Waste," New York Times, 3 May 94. The U.S. Supreme Court did not, in fact, rule that incinerator ash was toxic waste, as this article originally claimed. It only ruled that incinerator ash wasn't exempt from regulation as toxic waste under the provisions of the Resource Conservation and Recovery Act. The Environmental Protection Agency, under whose jurisdiction the regulation fell, issued no subsequent ruling that incinerator ash was toxic waste. See "Guidance for the Sampling and Analysis of Municipal Waste Combustion Ash for the Toxicity Characteristic," U.S. Environmental Protection Agency, Office of Solid Waste, EPA530-R-98-036, 1 Jun 95.

325. W. J. Deutsch, *Groundwater Geochemistry: Fundamentals and Application to Contamination* (CRC Press, 1997).

326. The metal ions responsible for most well water hardness are Ca^{2+}, Mg^{2+}, and Fe^{3+}, but there are others. Water hardness appears not to be well summarized in any official publication. The U.S. average mass fraction of 2.17×10^4 for $CaCO_3$ (8.6×10^5 of Ca^{2+}), widely quoted on the Internet, comes from J. C. Briggs and J. F. Ficke, "Quality of Rivers of the United States, 1975 Water Year—Based on the National Stream Quality Accounting Network (NASQAN)," U.S. Geological Survey Open-File Report 78-200, 1 Jan 77.

327. Well-documented cases of drinking water poisoning from landfill metals are probably nonexistent. See S. C. James, "Metals in Municipal Landfill Leachate and Their Health Effects," Am. J. Public Health **67**, 429 (1977); National Research Council, *Groundwater Contamination* (National Academies Press, 1984); C. D. Rail, ed. *Groundwater Contamination* (CRC Press, 2000); R. Metcalfe and C. A. Rochelle, "Chemical Containment of Waste in the Geosphere," Geological Society Special Publication No. 157 (Geological Society of London, 2000); H. M Selim and W. L. Kingery, *Geochemical and Hydrological Reactivity of Heavy Metals in Soils* (CRC Press, 2003); U. K. Singh *et al.*, "Assessment of the Impact of Landfill on Groundwater Quality: A Case Study of the Pirana Site in Western India," Environmental Monitoring and Assessment **141**, 309 (2008). By contrast, there are many cases of industrial contamination of groundwater by metals, notably from mines and metal works. See G. J. Barton, "Dissolved Cadmium, Zinc and Lead Loads from Ground-Water Seepage into the South Fork Coeur d'Alene River System, Northern Idaho, 1999," U.S. Geological Survey, Water-Resources Investigation Report 01-4274, 1 Jan 02; M. T. Koterba *et al.*, "Occurrence and Distribution of Mer-

cury in the Surficial Aquifer, Long Neck Peninsula, Sussex County, Delaware, 2003–04," U.S. Geological Survey Scientific Report 2006-5011, 1 Jan 06; B. Prasad and K. Sangita, "Heavy Metal Pollution Index of Ground Water of an Abandoned Open Cast Mine Filled with Fly Ash: A Case Study," Earth and Environ. Sci. **27**, 265 (2008). There are also many cases of metal contamination of groundwater from geological sources. See J. L. Barringer *et al.*, "Mercury in Ground Water, Septage, Leach-Field Effluent, and Soils in Residential Areas, New Jersey Coastal Plain," Science of the Total Environment **361**, 144 (2006); J. Nouri *et al.*, "Regional Distribution Pattern of Groundwater Heavy Metals Resulting from Agricultural Activities," Environ. Geol. **55**, 1337 (2008); C. Cartier, "Mercury Found in Coastal Groundwater," Santa Cruz News, 8 Jun 09. The mercury seepages at Moss Landing were only thirty miles from the famous New Almaden mercury mines, the first (and largest) mercury source in the continental United States, which played heavily in the California gold rush. See W. W. Bradley, *Quicksilver Resources of California with a Section on Metallurgy and Ore Dressing, Bulletin 78* (Calif. State Printing Office, 1918).

328. The metals required in greatest amounts in a healthy human diet are Fe, Zn, Mg, and Ca. See World Health Organization, *Vitamin and Mineral Requirements in Human Nutrition*, 2nd ed. (World Health Organization, 2004). Cu is required in trace amounts for proper neurological development. See I. E. Dreosti and R. M. Smith, eds., *Neurobiology of the Trace Elements* (Humana Press, 1983). Other metals are suspected of being important to health but not identified as such because there are no clear cases of deficiency-induced disease. Co is required in the human diet as a chemical component of vitamin B-12 but appears otherwise not to be necessary. Ni is required by some plants and animals but not by humans. Se and I, which are important in the diet, are often classified as metals, but they're actually nonmetals. As is a highly poisonous nonmetal.

329. L. Järup, "Hazards of Heavy Metal Contamination," Brit. Med. Bull. **68**, 167 (2003); I. C. Nnorom and O. Osibanjo, "Estimation of Consumption Emissions of Lead and Cadmium from Dry Cell Importation in Nigeria: 1980–1998," J. Appl. Sci. **6**, 1499 (2006); K. Nakamura, S. Kinoshita, and H. Takatsuki, "The Origin and Behavior of Lead, Cadmium and Antimony in MSW Incinerator," Waste Management **16**, 509 (1996).

330. The density of compacted incinerator ash lies between 1,700 kg/m^3 and 2,400 kg/m^3. See K. E. Forrester, "MSW-Ash Field Study: Achieving Optimal Disposal Characteristics," J. Environ. Eng. **116**, 880 (1990); R. W. Goodwin, *Combustion Ash/Residue Management: An Engineering Perspective* (William Andrew, 1994); R. Siddique, *Waste Materials and By-Products in Concrete* (Springer, 2007). Assuming that the big chunks of metal and glass have been removed for recycling, so that only 10% of remaining trash mass is ash (see note 316), the mass of ash generated by the United States in 2006 would occupy a volume of $0.1 \times 2.28 \times 10^{11}$ kg / (2,000 kg/m^3) $= 1.14 \times 10^7$ m^3, which is 8.6 cubic American football fields (see note 246).

331. S. Lyall, "Incinerator Ash in a New Guise: Building Blocks," New York Times, 7 Dec 89; T. Mangialardi, L. Piga, G. Schena, and P. Sirini, "Characteristics of MSW Incinerator Ash for Use in Concrete," Env. Eng. Sci. **15**, 291 (1998); H.-K. Cheong, J.-H. Tay, and K.-Y. Show, "Utilization of Municipal Solid Waste Fly Ash as Innovative Civil Engineering Materials," in *Use of Incinerator Ash*, R. K. Dhir, T. D. Dyer, and K. A. Paine, eds. (Thomas Telford, 2000); H. F. Hassan, "Recycling of Municipal Solid Waste Incinerator Ash in Hot-Mix Asphalt Concrete," Construction and Building Materials **19**, 91 (2005). Coal ash is more commonly used for these purposes at the moment, but mainly because it's in much greater supply. See O. E. Manz, "Worldwide Production of Coal Ash and Utilization in Concrete and Other Products," Fuel **76**, 691 (1997); M. Cheriaf, J. C. Rocha, and J. Péra, "Pozzolanic Properties of Pulverized Coal Combustion Bottom Ash," Cement and Concrete Research **29**, 1387 (1999); T. R. Naik *et al.*, "Cast-Concrete Products Made with FBC Ash and Wet-Collected Coal Ash," J. Mat. Civ. Eng. **17**, 659 (2005); J. Kavanagh, "Turning Toxic Coal Ash into Bridges, Buildings," CNN, 20 Mar 09.

332. Some marine dumping is probably not just thoughtlessness. See J. Perlez, "Report Heightens Pollution Dispute at Indonesian Bay," New York Times, 9 Nov 04. See also I. Peterson, "New Concern Raised by Waste-Dumping in Atlantic Off L.I.," New York Times, 31 Aug 87; S. Herrando-Pérez and C. L. J. Frid, "The Cessation of Long-Term Fly-Ash Dumping: Effects on Macrobenthos and Sediments," Marine Pollution Bull. **36**, 780 (1998).

333. The most relevant treaty is the Convention on the Prevention of Marine Pollution by Dumping of Wastes and Other Matter 1972 (London Dumping Convention) and the 1996 Protocol for That Convention (the London Protocol). See International Maritime Organization, *London Convention 1972 and 1996 Protocol* (International Maritime Organization, 2003). The relevant U.S. law is the 1972 Ocean Dumping Act. See C. Copeland, "Ocean Dumping Act: A Summary of the Law," CRS Report for Congress RS20028, 25 Aug 08.

334. The literature on metals in mineral springs is very large. Here are some recent publications: F. Queirolo *et al.*, "Total Arsenic, Lead, Cadmium, Copper, and Zinc in Some Salt Rivers in the Northern Andes of Antofagasta, Chile," Science of the Total Environment **255**, 85 (2000); J. W. Ball *et al.*, "Water-Chemistry Data for Selected Springs, Geysers and Streams in Yellowstone National Park, Wyoming 1999–2000," U.S. Geological Survey Open-File Report 02-382, 1 Jan 02; K. A. Hudson-Edwards and J. Archer, "Geochemistry of Arsenic in Spring and Stream Waters from San Antonio de los Cobres, NW Argentina," Mineralogical Mag. **72**, 425 (2008).

335. The concentration of trace metals in the sea is position-dependent. See F. J. Millero, *Chemical Oceanography* (CRC Press, 2005). For example, Cd is depleted in the upper 1 kilometer of the ocean but saturates at a mass fraction of about 1.0×10^{-9} moles/kg $\times$ 0.112 kg/mole = 1.12×10^{-10} at greater depths. The likely reason is scavenging by plankton, which then fall to the ocean bottom, decay, and release their Cd loads. Pb, by contrast, has mass fraction 1.50×10^{-10} moles/kg $\times$ 0.207

kg/mole = 3.11×10^{-11} in the upper 1 kilometer of the ocean but much smaller values at lower depths. The likely reason is that it gets constantly absorbed by the muds at the ocean bottom and then replenished through river runoff. The nominal mass fraction of Hg in the ocean is 1.0×10^{-11}, but Hg measurements reported in the literature vary by three orders of magnitude. At least some of this variation is due to location-dependent upwelling of Hg-laden water from springs. See M. Nishimura *et al.*, "Mercury Concentration in the Ocean," J. Oceanographical Soc. Jpn. **39**, 295 (1983); E. Nakayama *et al.*, "Chemical Analysis of Seawater for Trace Elements: Recent Progress in Japan on Clean Sampling Chemical Speciation of Trace Elements—A Review," Analytical Sciences **5**, 129 (1989); M. Sadiq, *Toxic Metal Chemistry in Marine Environments* (CRC Press, 1992).

336. The Pb mass fraction of municipal solid waste ash varies from very small to several hundred parts per million depending on the source. See T. L. Richard and P. B. Woodbury, "The Impact of Separation on Heavy Metal Contaminants in Municipal Solid Waste Composts," Biomass and Bioenergy **3**, 195 (1992); M. Aucott *et al.*, "Estimated Quantities and Trends of Cadmium, Lead and Mercury in U.S. Municipal Solid Waste Based on Analysis of Incinerator Ash," Water, Air, and Soil Pollution **206**, 349 (2010). Assuming that the representative value is 200 parts per million, the ash generated by burning all the trash generated in the United States in 2006 would contain roughly (see note 316) $0.2 \times 2.28 \times 10^{11}$ kg $\times 2.0 \times 10^{-4} = 9.12 \times 10^{6}$ kg of Pb. The average flow of the Amazon is 6.68×10^{12} m^3/y (see note 49). The flow of water that would be required to wash away this trash Pb and dilute it to the open ocean mass fraction (see note 335) of 3.1×10^{-11} is 9.12×10^{6} kg / ($3.1 \times 10^{-11} \times 6.68 \times 10^{12}$ m^3/y $\times 10^{3}$ kg/m^3) = 44.0 times the average flow of the Amazon.

337. The most far-reaching of these laws is the Mercury-Containing and Rechargable Battery Management Act, Public Law 104-142 (42 U.S. Code 14301-14336), May 13, 1996. It banned the sale in the United States of most batteries containing mercury and also banned the disposal of NiCd and sealed Pb-acid storage batteries in conventional trash. However, it made explicit exceptions for alkaline-manganese button batteries containing less than 25 milligrams of Hg and for Pb-acid batteries used in cars. A similar law (2006/66/EC), although with considerably weaker restrictions, was enacted by the European Parliament on September 6, 2006. See also Florida Battery Act, 403.7192 Florida Statutes, March 7, 1997; S. Chan, "Manhattan: Battery Disposal Is Banned," New York Times, 1 Dec 06. California went further and banned batteries of all kinds from its waste stream in 2001. See "California Universal Waste Rule," Department of Toxic Substances Control Number R-97-08. The degree to which California's law has been obeyed is not clear. Electronics recycling in the United States is presently mandated only at the state level. See R. Knutson, "Electronics Firms Fight Recycling Programs," Wall Street Journal, 2 Jul 09.

338. A case in point is the lead in present-day car batteries. The U.S. Environmental Protection Agency battery recycling website says (without attribution) that

(1) Americans purchase nearly 3 billion dry cell batteries every year, (2) nearly 99 million lead-acid car batteries are manufactured each year, (3) nearly 90% of lead-acid batteries are recycled, and (4) a typical lead-acid battery contains 60 to 80% recycled lead and plastic. Here is a more detailed accounting from "National Recycling Rate Study," Battery Council International, 1 Jul 03. The totals apply to the period 1997–2001 and are expressed in multiples of 10^9 kg of Pb: Battery Pb Consumed 5.502, Battery Pb Recycled −4.791, Battery Pb Scrap Import 0.128, Battery Pb Scrap Export −0.691. These numbers total to 0.148×10^9 kg during this period. This corresponds to 1.48×10^8 kg / 5.502×10^9 kg = 0.027 of Pb in circulation lost each year to the trash. The BCI report also gives the number of units sold as 4.73×10^8 batteries / (5 years) = 9.46×10^7 batteries per year. The disparity between this number and the EPA one is presumably due to unreported military quantities. A large potential error source in this accounting is the recycled Pb flow, which was assessed by surveys sent to secondary Pb smelters. The smelters typically measured their Pb by weighing delivery trucks before and after they discharged their loads and then multiplying the difference by a dimensionless factor to adjust for other substances (plastic) in the load. However, the 2–3% loss per year figure is corroborated by the following Pb wholesale flow figures from "Lead and Zinc Statistics," Monthly Bulletin of the International Lead and Zinc Study Group **46**, No. 2, February 2006, expressed as multiples of 10^9 kg/y: Secondary Pb Recovery 1.113, Ore Mined 0.439, Ore Imported 0.000, Ore Exported −0.295, Refined Metal Production −1.262. (The ore numbers refer to the Pb content of the ores, not the raw ore mass.) These figures sum to 0.013×10^9 kg/y. This may be an ore stockpile buildup or an accounting error, but it is small, in any event. The secondary Pb recovery figure in the LZSG report also agrees with the BCI report in that the fraction of the secondary Pb due to batteries, normally estimated at about 0.8, comes out to be $(4.79 + 0.128 − 0.691) \times 10^9$ kg/ (5 years $\times 1.127 \times 10^9$ kg/y) = 0.75. Adding the ore budget of the second report to the scrap flows of the first, we obtain $(0.439 + 0.000 − 0.295 + 0.026 − 0.138) \times 10^9$ kg/y = 0.032×10^9 kg/y for the lost Pb. This is 0.021 times the total Pb in circulation. See also A. M. Genaidy *et al.*, "An Exploratory Study of Lead Recovery in Lead-Acid Battery Lifecycle in US Market: An Evidence-Based Approach," Science of the Total Environment **407**, 7 (2008).

339. The U.S. government has shown extreme enthusiasm lately for electric cars. See S. Power and R. Smith, "A123 Looks Set to Land U.S. Funds for Battery," Wall Street Journal, 5 Aug 09; D. Hedgpeth and S. Wilson, "Grants Steered to Green Car Research," Washington Post, 6 Aug 09; T. Hamilton, "Lithium Battery Recycling Gets a Boost," Technology Review, 12 Aug 09; S. S. Carty, "States Eager to Power Up Electric Car-Battery Industry," USA Today, 21 Oct 09.

340. According to Tesla Motors, the Li-ion battery pack of its Roadster electric car has a mass of 450 kg, stores 53 kwh $\times 10^3$ w/kw $\times$ 3,600 sec/h = 1.91×10^8 joules, and gives the car a driving range of about 400 kilometers. See G. Berdichevsky *et al.*, "The Tesla Roadster Battery System," Tesla Motors, 16 Aug 06. Its

energy storage density is thus 1.91×10^8 joules / 450 kg = 4.2×10^5 joules/kg, a typical value for high-end Li-ion batteries (see note 127). Lead-acid batteries with the same storage capacity would have mass 1.91×10^8 joules / (1.26×10^5 joules/kg) = 1,516 kg. However, because the Roadster is a small vehicle, a more reasonable Li-ion battery mass for a typical future electric car would be 10^3 kg. According to the *Wall Street Journal*, customers in the United States bought about 15 million cars and light trucks in 2008 (just before the stock market crash). This figure does not include heavy trucks. Thus, if the entire U.S. automobile fleet were to become electric, and if the industry successfully recycled all but 1% of its Li-ion battery mass, the yearly flow of battery mass into the trash stream would be $0.01 \times 1.5 \times 10^7$ cars $\times 10^3$ kg/car = 1.5×10^8 kg. The U.S. Environmental Protection Agency says (without attribution) that 3 billion batteries presently make their way to the trash stream every year. A figure of 4 billion batteries, with total mass of 1.42×10^5 tons $\times$ 909 kg/ton = 1.29×10^8 kg, for the year 1992 is cited in D. J. Hurd, *Recycling of Consumer Dry Cell Batteries* (William Andrew, 1993). A sales figure of 5 billion batteries, with total mass 1.57×10^5 tons $\times$ 909 kg/ton = 1.48×10^8 kg, for the European market in the year 2000 is cited by the European Portable Battery Association (http://epbaeurope.net/statistics.html). The European and American markets are roughly the same size.

341. S. Romero, "In Bolivia, Untapped Bounty Meets Nationalism," New York Times, 2 Feb 09.

342. Lithium is typically administered as lithium carbonate or lithium citrate. Lithium salts sell under the trade names Cibalith, Eskalith, Lithane, and Lithobid.

343. Lithium does not accumulate in the body they way lead and mercury do. About 6 grams of it are required to kill a healthy adult. See M. R. Achong, P. G. Fernandez, and P. J. McLeod, "Fatal Self-Poisoning with Lithium Carbonate," Can. Med. Assn. J. **112**, 868 (1975). A comparable amount of excess potassium in the blood (10 millimolar) is also fatal. See C. Bacon, "Death from Accidental Potassium Poisoning in Childhood," British Med. J. **1**, 389 (1974). Lead, by contrast, does accumulate in the body. The concentration of it in the blood required to kill a person is 10 times smaller (1 millimolar). See C. L. Shrewsbury *et al.*, "Diagnosis of Poisoning of Beef Calves by Lead Paint," J. Anim. Sci. **4**, 20 (1945).

344. The battery of a typical gasoline automobile contains about 9 kilograms of Pb. The battery of a Toyota Prius contains about 15 kg of La, or 0.28 of the total battery pack mass. The electric motor of a Prius contains about 1 kg of Nd. See S. Gorman, "As Hybrid Cars Gobble Rare Earth Metals, Shortage Looms," Reuters, 31 Aug 09. See also L. Pietrelli *et al.*, "Rare Earths Recovery from NiMH Spent Batteries," Hydrometallurgy **66**, 135 (2002); D. A. Dertuol *et al.*, "Spent NiMH Batteries—Rare Earths Recovery and Leach Liquor through Selective Precipitation," Acta Metallurgica Slovaca **12**, 13 (2006).

345. Lanthanum is usually characterized as mildly toxic. See D. W. Bruce, B. E. Hietbrink, and K. P. DuBois, "The Acute Mammalian Toxicity of Rare Earth Nitrates

and Oxides," Toxicology and Applied Pharmacology **5**, 750 (1963); R. J. Palmer, J. L. Butenhoff, and J. B. Stevens, "Cytotoxicity of the Rare Earth Metals Cerium, Lanthanum and Neodymium in Vitro: Comparisons with Cadmium in a Pulmonary Macrophage Primary Culture System," Environ. Res. **43**, 142 (1987); S. Hirano and K. T. Suzuki, "Exposure, Metabolism and Toxicity of Rare Earths and Related Compounds," Environ. Health Perspect. **104** (Suppl. 1), 85 (1996); M. J. Barry and B. J. Meehan, "The Acute and Chronic Toxicity of Lanthanum to *Daphina carinata*," Chemosphere **41**, 1669 (2000).

CHAPTER 9: VIVA LAS VEGAS

346. Here are numbers comparing the three metropolitan areas: Los Angeles 792.0 12.8 15.3, Cairo 145.0 11.0 14.5, Karachi 78.0 13.1 11.8. The first two are the 2008 GDP expressed as a multiple of U.S.$ billion and the 2008 population expressed as a multiple of 10^6 from "UK Economic Outlook," PriceWaterhouseCoopers, November 2009. The third is the 2003 population expressed as a multiple of 10^6 from R. L. Forrestall, R. B. Greene, and J. B. Pick, "Which Are the Largest? Why Lists of Major Urban Areas Vary So Greatly," Tijdschrift voor Economische en Sociale Geografie **100**, 277 (2009).

347. P. Lorenz, D. Pinner, and T. Seitz, "The Economics of Solar Power," Forbes, 8 Jul 08; M. Halper, "BP's Sun King Reflects on Solar Power's Future," The Independent, 6 Dec 09. These articles need to be read with a grain of salt because they emphasize photovoltaic electricity and show a poor grasp of the solar concentrator side of the business, which is more profitable. However, the points they make about economies of scale apply to both kinds of plant.

348. The power flux from the sun just above earth's atmosphere is fixed by planck's constant h, Boltzmann's constant k_B, the speed of light c, the sun's temperature T, the sun's radius r, and the sun's distance from the earth R to be $(r/R)^2 \times 2\pi^5$ $(k_B T)^4 / (15\, h^3 c^2) = (6.96 \times 10^8\, m / 1.50 \times 10^{11}\, m)^2 \times 2\pi^5 \times (1.38 \times 10^{-23}\, joules/°K \times 5.78 \times 10^3\, °K)^4 / (15 \times (6.63 \times 10^{-34}\, joules\text{-}sec)^3 \times (3.0 \times 10^8\, m/sec)^2) = 1.36 \times 10^3$ watts/m^2. The ratio r/R is measured by the angle the sun subtends in the sky. The temperature T is measured by the wavelength peak of the solar light spectrum. The theoretical power flux agrees with the value measured experimentally by satellite. See R. C. Willson, "Total Solar Irradiance Trend During Solar Cycles 21 and 22," Science **26**, 1963 (1997).

349. The maximum solar flux received by the Sahara is slightly smaller than the ideal solar constant, about 930 watts/m^2, primarily because of atmospheric blockage of the sun's ultraviolet radiation (see note 348). The twenty-four-hour average of this flux (including twelve hours darkness) is 930 watts/m^2 / π = 296 watts/m^2. Multiplying by the area of the Sahara then gives 296 watts/m^2 × 9.0 × 10^{12} m^2 = 2.66×10^{15} watts (see note 200). The present energy needs of civilization are 1.4 × 10^{13} watts, or 0.53 percent of this number (see note 5). Enthusiasm in Europe for

solar energy from the Sahara continues to be very great. See F. Vorholz, "Kein Rückschlag: Thiemo Gropp, Verstand der Desertec-Stiftung, über den Fortgang des Wüstenstromprojekts," Die Zeit, 25 Mar 11; D. Wetzel, "Marokko Startet das Project Wüstenstrom," Die Welt, 24 Jul 10; D. Asendorpf, "Sonnenenergie aus Deutscher Hand," Die Zeit, 23 Jun 09; E. Kirschbaum, "Sahara Solar Plan 'Win-Win' for Europe and North Africa," Reuters, 24 Jun 09; J. Lubbadeh, "Wüstenstrom: Die Sonne über der Sahara Löst das Energieproblem," Der Spiegel, 17 Jun 09; L. Smith, "Solar Panels in the Sahara 'Could Power the Whole of Europe,'" London Times, 12 Mar 09; A. Jha, "Solar Power from Saharan Sun Could Provide Europe's Electricity, Says EU," The Guardian, 23 Jul 08; H. Kempf, "La Sahara Générateur d'Électricité?," Le Monde, 18 Nov 07.

350. The Andasol-1 plant in southern Spain provides a good test case of solar trough efficiency on account of being so well documented. The plant has an area of 2 km^2 and is designed to generate 50 megawatts for 7 hours. The average power is thus 5.0×10^7 watts $\times$ 7 hours / (24 hours) = 1.46×10^7 watts. The average insolation at the site is 2200 kwh/y $\times 10^3$ w/kw / (24 h/d $\times$ 365 d/y) = 250 watts/m^2. The overall plant efficiency is thus 1.45×10^7 watts / (250 watts/m^2 $\times$ 2.0×10^6 m^2) = 0.029. See "Parabolic Trough Power Plants Andasol 1 to 3," Solar Millennium AG, 1 Dec 08.

351. Here are the relevant characteristics of the four North American deserts: Mojave 0.65 0.65 5.7, Great Basin 4.11 4.11 4.9, Chihuahuan 4.55 1.20 5.7, Sonoran 3.12 1.00 5.7. The first number is the total desert area expressed as a multiple of 10^{11} m^2. The second number is the area of the desert lying inside the United States expressed as a multiple of 10^{11} m^2. Both of these come from a number of sources and have roughly 10% uncertainties that come from disagreement on what kind of land constitutes desert. The third number is a characteristic insolation expressed as a multiple of kilowatt-hours per square-meter-day, obtained from W. Marion and S. Wilcox, "Solar Radiation Data Manual for Flat-Plate and Concentrating Collectors," U.S. National Renewable Energy Laboratory, NREL/TP-463-5607, April 1994. They are, more precisely, the yearly averages for Las Vegas, NV; Ely, NV; El Paso, TX; and Tucson, AZ, respectively. Multiplying the second two numbers and adding gives 36.4. The total solar energy falling on the western deserts of the United States in one year is thus 36.4 $\times$ 10^{11} kwh/day $\times$ 3.6×10^6 joules/kwh $\times$ 365 days = 4.78×10^{21} joules. This is 53.5 times the 8.94×10^{19} joules of energy of all kinds consumed by the United States in 2006 (see note 309). If the solar energy were extracted by 3% efficient solar plants, this number would reduce to 53.5 $\times$ 0.03 = 1.60. Some of the land in question is city, mountain, park, and so forth, so the number is probably high.

352. The total solar power incident on the earth is 1.36×10^3 watts/m^2 $\times$ $2\pi \times (6.37 \times 10^6$ m)^2 = 6.93×10^{17} watts. According to the U.S. Energy Information Agency's International Energy Annual, the 2006 world electricity consumption was 1.64 $\times$ 10^{13} kwh/y $\times 10^3$ wh/kwh / (24 h/d $\times$ 365 d/y) = 1.87×10^{12} watts. Assuming that

half of this is being expended on earth's dark side and that half of the dark expenditure is used for light, the total solar power incident on the earth is $4 \times 3.47 \times 10^{17}$ watts / $(1.87 \times 10^{12}$ watts$) = 7.42 \times 10^{5}$ times the power of the civilization's lights on the dark side of earth.

353. One of the windiest places in the world is Antarctica—although, interestingly, not in the depths of winter. See J. Naithani *et al.*, "Analysis of Strong Wind Events Around Adelie Land, East Antarctica," Ann. Geophys. **46**, 385 (2003); M. Fischetti, "Turbines Spin in Antarctica," Scientific American, 2 Jul 09; M. L. Wald, "Catching the Wind in Antarctica," New York Times, 4 Nov 08; D. Murphy, *Rounding the Horn* (Phoenix Books, 2004).

354. D. L. Elliott *et al.*, "Wind Energy Resource Atlas of the United States," U.S. Solar Energy Research Institute, DOE/CH 10093-4, October 1986.

355. W. Pentland, "The Saudi Arabia of Solar Energy," Forbes, 22 Aug 08; A. Di-Paola, "Saudi Arabia to Target Solar Power in $100 Billion Energy Plan," Bloomberg, 31 Mar 11.

356. Here are some 2009 installment costs estimated by the California Energy Commission, expressed in 2009 U.S.$ per watt: Solar Parabolic Trough 3.69, Solar Parabolic Trough With Storage 5.41, Solar Photovoltaic 4.55, Onshore Wind 1.99, Offshore Wind 5.59, Combined Cycle Gas Turbine 1.04. See C. O'Donnell *et al.*, "Renewable Energy Cost of Generation Update," California Energy Commission, CEC-500-2009-084, August 2009.

357. Here are some figures from the International Energy Statistics Portal (www.iea.org/stats) for renewable electricity in 2008: Wind 218.504 78.66, Solar Photovoltaic 12.016 4.33, Solar Thermal 0.898 0.32. The first number is the gross electricity generation expressed as a multiple of 10^{12} watt-hours. The second number is the same quantity expressed as a multiple of 10^{16} joules (converted with 1 watt-hour = 3,600 joules). These statistics are roughly consistent with the ones from OECD countries published in *Renewables Information 2010* (OECD Press, 2010). The IEA portal figures for the United States alone in 2008 are Wind 55.696 20.05, Solar Photovoltaic 1.572 0.57, Solar Thermal 0.878 0.32. The U.S. Energy Information Agency's "Electric Power Annual 2009" gives these numbers: Wind 55.363 19.93, Solar (Thermal plus Photovoltaic) 0.864 0.31. Thus the IEA and EIA assess wind resources similarly but solar resources differently.

358. Here are some figures for the installed wind generation capacity of the world at the end of 2008, expressed as a multiple of 10^{9} watts: BP 122.1, IEA 120.5, WWEA 120.9. See the "Renewable Energy" section of the BP Statistical Review of World Energy, June 2010; "IEA Wind Energy Annual Report 2008," International Energy Agency, July 2009; "World Wind Energy Report 2010," World Wind Energy Association WWEA, April 2011. Had the turbines in question worked at full capacity all year long, they would have generated 1.2×10^{11} watts $\times$ 3,600 sec/h $\times$ 24 h/d $\times$ 365 d = 3.78×10^{18} joules. The amount of wind energy actually delivered was (see note 357) 7.87×10^{17} joules / $(3.78 \times 10^{18}$ joules$) = 0.21$ times this full-capacity number. The IEA report also cites 2008 figures for member countries (which

chiefly exclude China and India) of 91.8 gigawatts (2.90×10^{18} joules/year) of capacity and 194 terawatt-hours (6.98×10^{17} joules/year) of generation. The ratio of these numbers is 0.24. The WWEA report cites world 2010 figures of 196.6 gigawatts (6.20×10^{18} joules/year) and 430 terawatt-hours (1.55×10^{18} joules/year), which gives a ratio of 0.25. The U.S. Energy Information Administration's "Electric Power Annual" cites 2008 figures for the United States of 24.65 gigawatts (7.77×10^{17} joules/year) and 55.36 terawatt-hours (1.99×10^{17} joules/year), which gives a ratio of 0.26. The EIA's U.S. figures for 2009 are 34.30 gigawatts (1.08×10^{18} joules/year) and 73.89 terawatt-hours (2.66×10^{17} joules/year), which gives a ratio of 0.25.

359. For example, as it did in West Texas in 2008. See E. O'Grady, "Loss of Wind Causes Texas Power Grid Emergency," Reuters, 27 Feb 08; T. Fowler, "Wind Power Woes Not Sole Issue in Near Blackouts," Houston Chronicle, 1 Mar 08; P. Fairley, "Scheduling Wind Power," Technology Review, 17 Apr 08.

360. R. Gold, "Natural Gas Tilts at Windmills in Power Feud," Wall Street Journal, 2 Mar 10; C. Holly, "Texas Wind Boom Cutting into Fossil Generator Profits," Coal Power Magazine, 8 Oct 09.

361. K. Galbraith, "In Presidio, a Grasp at the Holy Grail of Energy Storage," New York Times, 6 Nov 10; P. Gupta, "Utility Wants to Deploy Largest Grid Battery Ever," Reuters, 26 Aug 09; P. Fairley, "Fixing the Power Grid," Technology Review, 17 Oct 07.

362. The U.S. Geological Survey's 2011 "Minerals Commodity Summary" estimates the total world supply of lithium to be 30 million tons. A lithium-ion cell has a potential of about 3.5 volts. The total amount of energy that could be stored in lithium-ion batteries made from all the world's estimated lithium is thus 3.5 volts $\times$ (3.0×10^7 tonnes $\times$ 10^3 kg/tonne / 0.006 kg/mole) $\times$ 6.022×10^{23} e/mole $\times$ 1.602×10^{-19} coulombs/e = 1.69×10^{18} joules. This is 1.69×10^{18} joules $\times$ 365 days/y / (4.4×10^{20} joules/y) = 1.40 days of the world energy use in 2006 (see note 5). The 2011 USGS estimate for the total world lead supply is 1.5 billion tons. Lead-acid cells have a potential of about 2.10 volts and involve the motion of two electrons per lead atom. The total amount of energy that could be stored in lead-acid batteries made from all the world's estimated lead is thus 2 $\times$ 2.10 volts $\times$ (1.5×10^9 tonnes $\times$ 10^3 kg/tonne / 0.207 kg/mole) $\times$ 6.022×10^{23} e/mole $\times$ 1.602×10^{-19} coulombs = 2.94×10^{18} joules.

363. D. Steward *et al.*, "Lifecycle Cost Analysis of Hydrogen Versus Other Technologies for Electrical Energy Storage," U.S. National Renewable Energy Laboratory, NREL/TP-560-46719, November 2009.

364. The solar power plants of the Andasol type (50 megawatts, 7.5 hours of thermal storage) (see note 368) presently in operation include Andasol-1, Andasol-2, Arcosol 50, Central Solar Termoelectrica La Forida, Extresol-1, Extresol-2, La Dehesa, and Manchasol-1. Plants of this type presently under construction include Andasol-3, Andasol-4, and Vallesol 50. Plants of this type under development include Extresol-3 and Manchasol-2. All of these plants are in Spain. Tower-type

plants using salt storage currently under development include Crescent Dunes and Rice, both of which are in the United States. See www.nrel.gov/csp/solarpaces.

365. The 2009 cost of pumped storage is measured by the design parameters of the Red Mountain Bar project in northern California: $\$1.3 \times 10^9 / (1.76 \times 10^{10}$ wh $\times 3,600$ sec/h) = $\$2.05 \times 10^{-5}$ joule^{-1}. See "Red Mountain Bar Pumped Storage Project," Turlock Irrigation District, 1 Sep 09. The 2009 cost of thermal storage is proprietary (and changing), but it is measured crudely by the capital cost difference between solar trough plants with storage and those without it. The Andasol 1 (50 megawatt, 2.0×10^6 m^2, 7.5 hours of storage) was capitalized at €311 in 2006 (see note 368). See "European Renewables Solar Deal of the Year 2006," Project Finance Magazine, 1 Mar 07. The Nevada Solar One plant (64 megawatt, 3.57×10^5 m^2, no storage) was capitalized in 2007 for $266 million. See "Acciona Energy Closes Long-Term Financing on Nevada Solar One," Business Wire, 30 Jul 07. Andasol-1 is load-leveled, whereas Nevada Solar One is not, so their costs are best pro-rated using the annual delivered energies, which is 134,000 megawatt-hours and 157,000 megawatt-hours, respectively (according to company documents). Assuming a 2006 exchange rate, the cost of storage is then roughly ($\$1.2$ €$^{-1} \times$ €$3.11 \times 10^8 - 1.57/1.34 \times \$2.66 \times 10^8) / (5.0 \times 10^7$ watts $\times 3,600$ sec/h $\times 7.5$h) = $\$4.56 \times 10^{-5}$ joule^{-1}. The California Energy Commission estimates the cost of thermal storage at ($\$5.41$ watt$^{-1} - \$4.55$ watt$^{-1}) / (3,600$ sec/h $\times 7.5$h) = $\$3.19 \times 10^{-5}$ joule^{-1} (see note 356). David Biello's *Scientific American* article quotes NREL engineer Greg Glatzmaier as estimating the Andasol storage cost at about $50 kwh$^{-1} / (10^3$w/kw $\times 3,600$ sec/h) = $\$1.39 \times 10^{-5}$joule^{-1} (see note 368). The formal 2006 NREL estimate was $\$3.95 \times 10^7 / (5.5 \times 10^7$ watts $\times 3,600$ sec/h $\times 9$h) = $\$2.22 \times 10^{-5}$ joule^{-1}. See B. Kelly and D. Kearney, "Thermal Storage Commercial Plant Design Study for a 2 Tank Indirect Molten Salt System," U.S. National Renewable Energy Laboratory, NREL/SR55040166, 1 Jul 06.

366. V. M. Boldyrev *et al.*, "Load-Following Nuclear Power Stations with Heat Accumulators," Atomic Energy **51**, 555 (1981); General Electric Company, "Power Generating Plant with Nuclear Reactor/Heat Storage System Combination," U.S. Patent 3,848,416, 19 Nov 74. Hydropumping eventually prevailed over thermal storage as a means of leveling nuclear loads (see note 201).

367. The first small prototype for molten salt solar energy storage was built at Sandia National Laboratory in 1983. See J. T. Holmes, "The Solar Molten Salt Electric Experiment," in *Alternative Energy Sources VI*, T. N. Veziroglu, ed. (Hemisphere Publishing Corp., 1985); R. Larson and R. E. West, eds., *Implementation of Solar Thermal Technology* (MIT Press, 1996). It was subsequently tested at the 10-megawatt level as the 1996 Solar Two upgrade to the Solar One experiment in Barstow. See C. E. Tyner, J. P. Sutherland, and W. R. Gould Jr., "Solar Two: A Molten Salt Power Tower Demonstration," Sandia National Laboratory, SAND-95-1828c, 1 Jan 95; "Solar Two," U.S. Department of Energy DOE/GO-10098-562, 1 Apr 98.

368. The world's first commercial solar thermal power plant with molten salt storage, Andasol-1, went on line in Spain in November 2008 (see note 350). Its sis-

ter plants, Andasol-2 and Andasol-3, went into service in June 2009 and June 2011, respectively. Each Andasol plant has an area of 2 square kilometers, is rated at 50 megawatts peak power, and is equipped with enough storage to sustain this power for 7.5 hours in the dark. See "Andalusiens Sonne Ernten," Die Zeit, 22 Oct 08; D. Biello, "How to Use Solar Energy at Night," Scientific American, 18 Feb 09; M. Crotty, "Interview with Rainer Aringhoff," Beyond Zero Emissions, 14 May 09; M. Koerth-Baker, "Forget Lightning. How Do We Catch Sunshine in a Bottle?," Discover Magazine, 17 Jun 09; "Parabolic Trough Power Plants Andasol 1 to 3," Solar Millennium AG, 1 Dec 08.

369. M. Kanellos, "Full Steam Ahead for Nevada Solar Project," CNET News, 12 Mar 07. Andasol-1 had broken ground on 3 Jul 06.

370. P. Fairley, "Storing Solar Power Efficiently," Technology Review, 27 Sep 07.

371. J. L. Lunsford, "Solar Venture Will Draw on Molten Salt," Wall Street Journal, 2 Jan 08.

372. M. Wald, "New Ways to Store Solar Energy for Nighttime and Cloudy Days," New York Times, 15 Apr 08.

373. P. Pae, "A Solar Plant That's Worth Its Salt," Los Angeles Times, 29 May 09.

374. C. Lombardi, "Solar Plant to Store Energy in Molten Salt," CNET News, 18 Nov 09.

375. "SoCal Desert Highway Landmark Demolished," San Jose Mercury News, 25 Nov 09.

376. T. Woody, "Generating Solar Power After Dark," New York Times, 23 Dec 09.

377. The sodium-potassium nitrate eutectic occurs at 50 molar % $NaNO_3$ and 50 molar % KNO_3, where the melting temperature is 221°C. Binary solar salt (66 molar % $NaNO_3$ and 34 molar % KNO_3 has a melting temperature of 238°C. The melting temperatures of pure $NaNO_3$ and pure KNO_3 are 307°C and 334°C, respectively. See R. W. Bradshaw and N. P. Siegel, "Molten Nitrate Salt Development for Thermal Energy Storage in Parabolic Trough Solar Power Systems," in *Proc. of the 2nd Intl. Conf. on Energy Sustainability, Vol. II* (Am. Soc. Mech. Eng., 2009); "Density of Molten Elements and Representative Salts," D. R. Lide, *Handbook of Chemistry and Physics*, 84th ed. (CRC Press, 2003); G. J. Janz, "Thermodynamic and Transport Properties of Molten Salts: Correlations Equations for Critically Evaluated Density, Surface Tension, Electrical Conductance, and Viscosity Data," J. Phys. Chem. Ref. Data **17** (Suppl. 2), 1 (1988). Salts expand significantly upon melting. See K. S. Pitzer, "Thermodynamics of Ionic Fluids Over Wide Ranges of Temperature," Pure and Appl. Chem **59**, 1 (1987).

378. The melting point of eutectic solder (63 weight % Sn and 37 weight % Pb) is 183°C. The melting points of pure Sn and Pb are 232°C and 327°C, respectively.

379. S. H. Goods and R. W. Bradshaw, "Corrosion of Stainless Steels and Carbon Steel in Molten Mixtures of Industrial Nitrates," J. Mat. Eng. Performance **13**, 78 (2004); D. C. Smith *et al.*, "Operation of Large-Scale Pumps and Valves in Molten Salt," J. Solar Eng. **116**, 137 (1994); A. A. Attia *et al.* "Corrosion Behavior of Stainless

Steel Alloys in Molten (Na,K) Eutectic Mixtures," Materialwissenschaft u. Werk-stofftechnik **30**, 559 (1999).

380. A standard (small) Olympic pool has volume 50 m × 25 m × 2 m = 2,500 m^3. The heat capacity of water is 4.19 joules $^{\circ}K^{-1}cm^{-3}$. The energy required to heat the Olympic pool from 0°C to 100°C is thus 4.19 × 10^6 joules $^{\circ}K^{-1}m^{-3}$ × 2,500 m^3 × 100°K = 1.05 × 10^{12} joules. The time it would take a one-kilowatt toaster to do this heating is 1.05 × 10^{12} joules / (1,000 watts × 3,600 sec/h × 24 h/d × 365 d/y) = 33.3 years. Replacing the pool with a cube 100 meters on a side gives an energy content of 4.18 × 10^{14} joules and a toaster heating time of 13,250 years.

381. According to California's Energy Consumption Data Management website (http://ecdms.energy.ca.gov), the total 2007 electricity energy usages for the five counties making up the Los Angeles, San Bernardino, and Ventura metropolitan areas were Los Angeles 7.08 2.55, Orange 2.09 0.75, Riverside 1.46 0.53, San Bernardino 1.47 0.53, and Ventura 0.57 0.21. The first number is the total usage expressed as a multiple of 10^{13} watt-hours. The second is the total usage expressed as a multiple of 10^{17} joules (1 watt-hour = 3,600 joules). The total average power demand of Greater Los Angeles in 2007 was thus (2.55 + 0.75 + 0.53 + 0.53 + 0.21) × 10^{17} joules/y / (3,600 sec/h × 24 h/d × 365 d/y) = 1.45 × 10^{10} watts, or about the same as greater New York (see note 154). The energy in 100 cubic meters of boiling water would thus power Greater Los Angeles for 4.18 × 10^{14} joules / (1.45 × 10^{10} watts × 3,600 sec/h) = 8.01 hours (see note 380).

382. The specific gravity of molten solar salt is about 1.9 (see note 377). Its average atomic mass per atom is (0.66 × 23 + 0.34 × 39 + 14 + 3 × 16) / 5 = 18.1. The heat capacity per unit volume is thus 1900 kg/m^3 × 3 × 8.314 joules mole^{-1} $^{\circ}K^{-1}$ / (0.0181 kg/mole) = 2.62 × 10^6 joules $^{\circ}K^{-1}m^{-3}$, or 0.63 times that of water (see note 380).

383. Oil refinery tanks have heights up to 18 meters and diameters up to 66 meters. See "Welded Steel Tanks for Oil Storage," American Petroleum Institute, API Standard 650, November 1998. (The Andasol-1 tanks are 14 meters high.) (See note 368.) Assuming the upper limit of this range, a field 1 kilometer on each side packed with round tanks filled with molten solar salt (see note 382) would power Greater Los Angeles (see note 381) for $\pi/12^{1/2}$ × 18 m × (1,000 m)^2 × 2.62 × 10^6 joules $^{\circ}K^{-1}m^{-3}$ × 100°K / (1.45 × 10^{10} watts × 3,600 sec/h × 24 h/d) = 3.41 days.

384. D. Hull, "Energy Department Announces $2.1 Billion Loan Guarantee for Mojave Solar Project," San Jose Mercury News, 18 Apr 11; D. R. Baker, "U.S., Google Back BrightSource Mojave Solar Plants," San Francisco Chronicle, 12 Apr 11; M. Richardson, "Solar Boom in the American Desert," Japan Times, 12 Nov 10; L. Sahagun, "Renewable Energy Sparks a Probe of a Modern-Day Land Rush," Los Angeles Times, 1 Jun 09; P. Foy, "Alternate-Energy Scramble on Across West," San Francisco Chronicle, 28 Sep 09; D. R. Baker, "Bechtel to Build Solar Plants for BrightSource," San Francisco Chronicle, 9 Sep 09; S. Kraemer, "Ex-United Technologies Rocket Scientists to Build 150 MW Solar Heliostat in Sonoran Desert," Scientific American, 4 Nov 09; R. Weiss, "Siemens to Buy Solel for $418 Million to

Expand Solar Business," Bloomberg, 15 Oct 09; L. Isensee, "U.S. Army to Build 500 MW Solar Power Plant," Reuters, 15 Oct 09; S. Patel, "Interest in Solar Tower Technology Rising," Power Magazine, 1 May 09; A. C. Revkin, "California Utility Looks to Mojave Desert Project for Solar Power," New York Times, 11 Feb 09; G. Chang, "Google, Chevron Build Mirrors in Desert to Beat Coal with Solar," Bloomberg, 23 May 08.

385. Here is a partial list of the more famous plants of the early twenty-first-century Mojave desert solar boom, with their design powers expressed as a multiple of 10^9 watts: Abengoa Mojave Solar 0.25, Acciona Fort Irwin 0.50, Bright Source Coyote Springs 0.40, Bright Source Ivanpah 0.37, Bright Source PG&E 0.6, First Solar Desert Sunlight 0.55, Harper Lake Solar 0.16, Siemens Mojave 0.50, Solar Millenium Amargosa 0.50, Solar Millennium Blythe 1.00, Solar Millennium Palen 0.50, Solar Reserve Crescent Dunes 0.10, Solar Reserve Rice 0.15, Next Extra Genesis 0.50, NRG Gaskell Sun Tower 0.24. As of April 2011, the California solar project applications approved or pending at the U.S. Bureau of Land Management totaled 1.45×10^{10} watts and involved 1.53×10^5 acres $\times 4.446 \times 10^3$ m^2/acre = 2.20×10^8 m^2 of land (www.blm.gov/ca/st/en/fo/cdd/alternative_energy/SolarEnergy.html). The solar projects approved and pending at the California Energy Commission totaled 4.49×10^9 watts (www.energy.ca.gov/siting/solar).

386. The average 2010 wholesale price of electricity in California was $40 per megawatt-hour ($1.11 \times 10^{-8}$ joule^{-1}). See "Market Issues and Performance Annual Report," California Independent System Operator, April 2011. The retail cost of electricity at this time was roughly $100 per megawatt-hour in 2010. According to the California Energy Commission, the 2009 levelized costs of wind and solar trough electricity were about $70 and $250 per megawatt-hour, respectively. See J. Klein, "Comparative Costs of California Central Station Electricity Generation," California Energy Commission CEC-200-2009-017-SF, January 2010. The solar trough number is cross-checked by the published feed-in tariffs for Andasol-1, which were €270 ($340) per megawatt-hour in 2008. There is good reason to trust this number because the tariff was renegotiated upward with the Spanish government while Andasol-1 was being built, presumably in response to a sharpening understanding of costs. The CEC report unfortunately has complex adjustments for future regulations and tariffs. Its natural gas generation costs, for example, are more than twice the actual wholesale cost of electricity in 2010. However, both California's actual wholesale electricity price and CEC's estimate for wind electricity costs are consistent with international values published in *Projected Costs of Generating Electricity* (OECD Press, 2010).

387. N. Cummings, "Tehachapi: Planned for Prosperity," Wind Systems Magazine, August 2010; L. Corum, "New Wind: Opening up California's Tehachapi Pass," Refocus **6**, 26 (2005).

388. The Mojave has an area of 6.5×10^{10} m^2 and an average insolation of 5.7 kilowatt-hours per square meter per day (see note 351). Assuming 3% conversion

efficiency, the entire Mojave thus has the potential to generate $0.03 \times 6.5 \times 10^{10}$ m^2 $\times$ 5.7 kwh/d $\times 10^3$ w/kw / (24 h/day) = 4.63×10^{11} watts.

389. See L. Riddell, "Bay Area Solar Companies Get Scorched," San Francisco Business Times, 6 Feb 09; P. Voosen, "Spain's Solar Market Crash Offers a Cautionary Tale about Feed-In Tariffs," New York Times, 18 Aug 09; C. Webb, "The Pain in Spain Falls Mainly on the Plain," Power Engineering International, 13 Nov 09; E. Rosenthal, "Solar Industry Learns Lessons in Spanish Sun," New York Times, 8 Mar 10.

390. See P. Lorenz, D. Pinner, and T. Seitz, "The Economics of Solar Power," Forbes, 9 Jul 08; M. Halper, "BP's Sun King Reflects on Solar Power's Future," The Independent, 6 Dec 09. These articles need to be read with a grain of salt, for they emphasize photovoltaic electricity and show a poor grasp of the solar concentrator side of the business, which is more profitable. However, the points they make about economies of scale apply to both kinds of plant.

391. D. Biello, "Sunny Outlook: Can Sunshine Provide All of U.S. Electricity?," Scientific American, 19 Sep 07; K. Zweibel, J. Mason, and V. Fthenakis, "A Solar Grand Plan," Scientific American, 1 Jan 08.

392. The engineering cost of electricity made from nuclear reactors is a politically charged subject, with many people arguing (with numbers) that it's expensive and many other arguing (with other numbers) that it's cheap. The California Energy Commission is in the former category and the International Energy Agency is in the other (see note 386). But the fact that companies are still talking about building nuclear plants, notwithstanding their immense unpopularity, proves that the immediate costs of nuclear electricity are less than those of natural gas electricity. However, the long-term costs of nuclear security and waste handling aren't known. Many people feel that these costs are infinite and that nuclear energy is thus the most expensive energy in the world. See A. Caputo, "Nuclear Could Still Edge Out Gas," San Antonio Express-News, 14 Dec 09; "No New Nukes—Plants, That Is," Los Angeles Times, 28 Nov 09.

393. D. B. Guenther, "Age of the Sun," Astrophys. J. **339**, 1156 (1989).

394. J. C. Wheeler, *Cosmic Catastrophes: Exploding Stars, Black Holes and Mapping the Universe* (Cambridge U. Press, 2007).

395. By remarkable coincidence, the sun and the moon subtend the same solid angle of sky. Thus, if the moon's surface temperature were the same as the sun's, it would look like the sun to an earth observer (see note 348).

396. Bricks have a specific gravity of about 1.84 and are made chiefly of Al_2O_3, the average atomic weight of which is $(2 \times 27 + 3 \times 16) / 5 = 20.4$. Their heat capacity per unit volume is thus 1840 kg/m^3 $\times$ 3 $\times$ 8.312 joules °K^{-1}mole^{-1} / (0.0204 kg/mole) = 2.25×10^6 joules °K^{-1}m^{-3}. Assuming that a layer of bricks 0.1 m thick is well insulated on the bottom, so that it loses heat only radiatively from the top surface, and that its initial temperature is the sun's (T = 5778 °K), its cooling time is roughly 2.25×10^6 joules °K^{-1}m^{-3} $\times$ 0.1 m / (5.67×10^{-8} watts °K^{-4}m^{-2} $\times$ (5778 °K)^3) = 206 seconds (see note 395).

397. This is partially because lightbulbs are designed to minimize nonradiative heat loss. They work by heating the filament hotter and hotter until it glows. Any nonradiative heat loss would thus increase the amount of electricity we must expend to produce a given amount of light, so it's minimized in the design. Lightbulb manufacturers would like to get rid of the infrared radiative losses as well, but they can't when the light-producing effect is incandescence.

398. The three reactors at the Palo Verde nuclear plant, each of which produces 3.8×19^9 watts of heat, would be considered large. According to the SmartSignal Corporation, makers of reactor monitoring equipment, each main coolant pump at Palo Verde delivers 114,625 gallons per minute through thirty-inch pipes. See "Early Warning of Reactor Pump Seal Degradation," Application Note Number 6, SmartSignal Corporation, 2003. The total area through which the reactor's power is pumped is thus $4 \times \pi \times (15.0 \text{ in } / 40.8 \text{ in/m})^2 = 1.82 \text{ m}^2$. Home-entry door dimensions vary, but they're typically $1 \text{ m} \times 2 \text{ m} = 2 \text{ m}^2$. To put this pumping feat in perspective, one Palo Verde pump produces a flow 114,625 gals/min / (1,500 gals/min) = 76.4 times that of the largest fire hydrants.

399. Assuming a maximum midday insolation of 930 watts per square meter appropriate for the central Sahara (see note 349), the thermal power of one Palo Verde reactor is spread out over area 3.9×10^9 watts / (930 watts/m^2) = 4.19×10^6 m^2 (see note 398). The number of Andasol-type plants (20 megawatts average power, 2 square kilometers) (see note 368) required to deliver the same amount of power as one Palo Verde reactor (36% thermal efficiency) is $0.36 \times 3.9 \times 10^9$ watts / $(2.0 \times 10^7$ watts) = 70.2. The amount of land they would take up is $70.2 \times 2.0 \times 10^6$ m^2 = 1.40×10^8 m^2.

400. C. Sullivan, "U.S. Halts Mojave Solar Project Over Species Concerns," New York Times, 28 Apr 11; T. Woody, "Solar Energy Faces Tests on Greenness," New York Times, 23 Feb 11; M. Rafferty, "Tribe Sues to Block Desert Solar Project," East County Magazine, 17 Nov 10; S. Tavares, "Amargosa Valley Warms Up to Solar Plan," Las Vegas Sun, 21 Jan 10; T. Woody, "Desert Vistas vs. Solar Power," New York Times, 21 Dec 09; M. Waite, "Comment Period Opens for Solar Project Near Tonopah," Pahrump Valley Times, 9 Dec 09; M. Waite, "Solar Millennium to Use Dry Cooling," Pahrump Valley Times, 18 Nov 09; T. Woody, "Alternative Energy Projects Stumble on a Need for Water," New York Times, 29 Sep 09; L. Sahagun, "Solar Energy Firm Drops Plan for Project in Mojave Desert," Los Angeles Times, 18 Sep 09; R. Glennon, "Is Solar Power Dead in the Water?," Washington Post, 7 Jun 09; "Amargosa Desert: 'Worthless Habitat'?," Basin and Range Watch, 14 May 09; J. Eilperin and S. Mufson, "Renewable Energy's Environmental Paradox," Washington Post, 16 Apr 09; R. Simon, "Feinstein Wants Desert Swath Off-Limits to Solar, Wind," Los Angeles Times, 25 Mar 09; M. Kettmann and C. Plains, "Solar Power: Eco-Friendly or Eco-Blight?," Time, 24 Mar 09; B. M. Pavlik, "Could Green Kill the Desert?," Los Angeles Times, 15 Feb 09; D. Frosch, "U.S. Lifts Moratorium on New Solar Projects," New York Times, 3 Jul 08; D. Frosch, "Citing Need for Assessments, U.S. Freezes Solar Energy Projects," New York Times, 27 Jun 08; P. Maloney, "Solar Projects Draw New Opposition," New York Times, 23 Sep 08.

CHAPTER 10: UNDER THE SEA

401. One knows this because the heat flux from the earth's interior, which averages about 75 milliwatts/m^2 at the ocean floor, would heat the ocean to its surface temperature in 4,000 m × 4.18 × 10^6 joules °K^{-1}m^{-3} × 20 °K / (0.075 watts/m^2 × 3,600 sec/h × 24 h/d × 365 d/y) = 1.42 × 10^5 years if the cold water were not constantly replenished. The cold water is not just left over from ice age glacier melt, because that melt raised the ocean surface only by 150 meters (see note 46). Geothermal heat would have brought that amount of water up to surface temperatures in 5.3 × 10^3 years. See D. S. Chapman, "Thermal Gradients in the Continental Crust," Geolog. Soc. London, Special Publications **24**, 63 (1986); R. E. Blackett, "Geothermal Gradient Data for Utah," Utah Geol. Survey, 1 Feb 04; A. G. Whittington, A. M. Hofmeister, and P. I. Nabelek, "Temperature-Dependent Thermal Diffusivity of the Earth's Crust and Implication for Magnetism," Nature **458**, 319 (2009).

402. The extinction depth of sunlight varies with circumstances but is about 40 meters in the deep ocean. See J. Babson *et al.*, "Cosmic-Ray Muons in the Deep Ocean," Phys. Rev. D **42**, 3613 (1990). Several sources identify the light intensity from the moonless night sky (airglow) as 10^{-8} times that of the midday sun. (The light of the stars is about 10 times smaller.) Thus the ocean is already black as night at depth 8 ln (10) × 40 m = 737 meters. The intensity of midday sun penetrating to the ocean bottom is exp(−4,000 m/40 m) = 3.7 × 10^{-44}. This number is so small that it's swamped by Cerenkov light generated by cosmic-ray muons and β decays of ^{40}K dissolved in seawater.

403. The pressure at the ocean's abyssal plain is 1,000 kg/m^3 × 9.8 m/sec^2 × 4,000 m = 3.92 × 10^7 pascals (387 atmospheres) (see note 111). This exceeds the critical pressure of 2.2 × 10^7 pascals, so water won't boil (see note 123). This is how undersea hydrothermal vents can issue water at 400°C with no boiling bubbles. See D. Coumou, T. Driesner, and C. A. Heinrich, "The Structure and Dynamics of Mid-Ocean Ridge Hydrothermal Systems," Science **321**, 1825 (2008).

404. National Research Council, *Our Seabed Frontier* (National Academies Press, 1989).

405. The kind of trouble we're talking about is British Petroleum's 2010 Macondo Well disaster at the bottom of the Gulf of Mexico, which gushed uncapped for 86 days and spilled 5 billion bbl of oil. See J. Achenbach, *A Hole at the Bottom of the Sea: The Race to Kill the BP Oil Gusher* (Simon and Shuster, 2011); C. Dreifus, "Revisiting the Deepwater Horizon Oil Spill," New York Times, 21 Mar 11. Nonetheless, the shift of emphasis toward deepwater oil exploration is already under way. See S. Hargreaves, "Oil's Future Is in Deepwater Drilling," CNN Money, 11 Jan 11; L. Gray, "MPs Give Backing to Deep Water Drilling Off UK," The Telegraph, 3 Jan 11; P. Baker and J. M. Broder, "White House Lifts Ban on Deepwater Drilling," New York Times, 12 Oct 10; J. Mouawad and B. Meier, "Risk-Taking Rises as Oil Rigs in Gulf Drill Deeper," New York Times, 29 Aug 10; S. Mufson, "Trend Toward Deep-

Water Drilling Likely to Continue," Washington Post, 22 Jun 10; T. Murphy, "Brazil's OGX Estimates Campos Oil Find at 1–2 Billion Barrels," Wall Street Journal, 22 Dec 09; J. Fick, "Brazil Petrobras: Iara Formation Tests Confirm Oil Estimates," Wall Street Journal, 9 Dec 09; S. Wrobel, "Huge Gas Reserves Discovered Off Haifa," Jerusalem Post, 19 Jan 09; J. Ryall, "Japan and China in Dispute Over Undersea Oil Fields," The Telegraph, 5 Jan 09; J. Carroll and C. Caminada, "Petrobras Discovers World's Third-Largest Oil Field," Bloomberg, 15 Apr 08; A. Evans-Pritchard, "Oil Crisis Triggers Fevered Scramble for the World's Seabed," The Telegraph, 27 May 08. The most poignant example is probably the redirection of financial resources from exhausting North Sea fields to more promising Canadian and West African ones. See A. Beckett, "Deeper, Rougher, Further—in Search of the Last North Sea Oil," The Guardian, 27 Oct 07; E. Mearns, "North Sea Petroleum Reserves," The Oil Drum: Europe, 5 Oct 09; T. Webb, "North Sea Oil's Ebbing Tide," The Guardian, 4 May 08.

406. J. Forero, "Brazil Girds for Massive Offshore Oil Extraction," Washington Post, 7 Dec 09; Y. Rosen, "Shell Alaska Chukchi Drill Plan Has Conditional OK," Reuters, 7 Dec 09; M. Stigset, "Arctic Oil Tempts Norway to Seek Drilling at 'Gates of Hell,'" Bloomberg, 24 Sep 09; T. Halpin, "Russia Warns of War within a Decade Over Arctic Oil and Gas Riches," London Times, 14 May 09.

407. Methane hydrate ice is stable under pressure to temperatures as high as 18°C (see note 56). Estimates of the amount of such ice under the sea have been declining but are still greater than the conventional methane reserves reported in the BP Statistical Review of World Energy 2008. See G. R. Dickens, C. K. Paull, and P. Wallace, "Direct Measurement of in Situ Methane Quantities in a Large Gas-Hydrate Reservoir," Nature **385**, 426 (1997); A. V. Milkov, "Global Estimates of Hydrate-Bound Gas in Marine Sediments: How Much Is Really Out There?," Earth-Sci. Rev. **66**, 183 (2004).

408. Some spectacular videos of ROVs in action in the deep ocean are posted at Oceaneering International's website. See also T. Zeller Jr., "Will Robots Clean Up Future Oil Spills?," New York Times, 24 Aug 10; L. Greenemeier, "Gulf Oil Spill Highlights the Increasing Dependence on Deep-Sea Robots," Scientific American, 7 May 10; I. Yarett, "Our Eyes Underwater," Newsweek, 19 May 10; *The World ROV Market Report 2010–2014* (Douglas-Westwood, 2009); J. K. Borchardt, "Robots of the Sea," Mechanical Engineering Magazine **130**, No. 7, 36 (2008); M. S. Lavine, D. Voss, and R. Coontz, "A Robotic Future," Science **318**, 1083 (2007); "How Deep-Water Drilling Works," Forbes, 4 Oct 04; *Undersea Vehicles and National Needs* (National Academies Press, 1996); W. J. Broad, "Undersea Robots Open a New Age of Exploration," New York Times, 13 Nov 90.

409. High-end flight simulation suitable for training military and commercial pilots requires computational resources far exceeding those of a home computer. See J. M. Rolfe and K. J. Staples, eds., *Flight Simulation* (Cambridge U. Press, 1988); G. L. Waltman, "Black Magic and Gremlins: Analog Flight Simulations at NASA's

Flight Research Center," U.S. National Aeronautics and Space Administration, NASA SP-2000-4520 (2000); S. Van Hoek and M. C. Link, *From Sky to Sea: A Story of Edwin A. Link* (Best Publishing, 2003); D. Allerton, *Principles of Flight Simulation* (Wiley, 2009).

410. "Underwater Robot Makes History Crossing Gulf Stream," Science Daily, 8 Nov 04; E. Fiorelli *et al.*, "Multi-AUV Control and Adaptive Sampling in Monterey Bay," IEEE J. of Oceanic Engineering **31**, 935 (2006); J. Fildes, "Robot Glider Harvests Ocean Heat," BBC News, 8 Feb 08; D. Brown, "Submersible Glider Spent Months Collecting Data on Atlantic Waters," Washington Post, 15 Dec 09; "Fleet of High-Tech Robot 'Gliders' to Explore Oceans," Science Daily, 18 Jan 10.

411. The biggest deep sea robots at the moment are pipe trenchers such as Canyon Offshore's T750-1 Supertrencher (23 tonnes), CTC Marine's UT-1 Ultra Trencher (63 tonnes), and CTC Marine's RT-1 Rock Trencher (202 tonnes).

412. B. Uttal, "A Respite in the Chip Price War," Fortune, 12 May 86.

413. At the moment, the only "underwater hotel" in the world with such air locks is Jules' Undersea Lodge in Key Largo, Florida. Its floor is only 6.4 meters below the ocean surface, presumably because the overpressure required to stabilize its moon pool entrance is 0.6 atmosphere. The roof is only 3.2 meters below the surface (the lodge website says 5.8 meters). See A. Onion, "Five-Star Hotel Under the Sea," ABC News, 10 Feb 05; M. Behar, "1,200 Square Feet Under the Sea," Popular Science, 12 Dec 06.

414. R. D. Hof, "My Virtual Life," Business Week, 1 May 06.

415. Robotic mining of the ocean isn't that futuristic, actually. See "Undersea Metal Miners Prepare to Go in Deep," The Chief Engineer, 31 Jan 10.

416. An up-to-date list of known undersea hydrothermal vents (about 550 in all) and a map showing their locations is maintained by the Pacific Marine Environment Laboratory (www.pmel.noaa.gov/vents). Current estimates are that about 70% of the earth's geothermal heat production occurs at sea floor spreading boundaries. See J. G. Sclater, B. Parsons, and C. Jaupart, "Oceans and Continents: Similarities and Differences in the Mechanism of Heat Loss," J. Geophys. Res. **86** (B12), 11535 (1981).

417. The vent temperatures of black smokers are typically 350° C. The total power is difficult to measure, but estimates for the whole field are typically between 0.3 and 1.0 gigawatt. See E. T. Baker, "Hydrothermal Cooling of Midocean Ridge Axes: Do Measured and Modeled Heat Fluxes Agree?," Earth Planet. Sci. Lett. **263**, 140 (2007); P. Ramondenc *et al.*, "The First Measurements of Hydrothermal Heat Output at 9° 50' N, East Pacific Rise," Earth Planet. Sci. Lett. **245**, 487 (2006); H. N. Edmonds *et al.*, "Discovery of Abundant Hydrothermal Venting on the Ultraslow-Spreading Gakkel Ridge in the Arctic Ocean," Nature **421**, 252 (2003); D. R. Yoerger, D. S. Kelley, and J. R. Delaney, "Fine-Scale Three-Dimensional Mapping of a Deep-Sea Hydrothermal Vent Site Using the Jason ROV System," Intl. J. Robotics Res. **19**, 1000 (2000); P. Anshutz and G. Blanc, "Heat and Salt Fluxes in the Atlantis II Deep (Red Sea)," Earth Planet. Sci. Lett. **142**, 147 (1996); K. L. Von Damm,

"Seafloor Hydrothermal Activity: Black Smoker Chemistry and Chimneys," Annu. Rev. Earth Planet. Sci. **18**, 173 (1990); D. R. Converse, H. D. Holland, and J. M. Edmond, "Flow Rates in the Axial Hot Springs of the East Pacific Rise (21°N): Implications for the Heat Budget and the Formation of Massive Sulfide Deposits," Earth Planet. Sci. Lett. **69**, 159 (1984); E. T. Baker and G. J. Massoth, "Hydrothermal Plume Measurements: A Regional Perspective," Science **234**, 980 (1986).

418. Estimates for the total hydrothermal energy flux of the midocean ridges are typically 10^{13} watts × 3,600 sec/h × 24 h/d × 365 d/y = 3.15×10^{20} joules per year, or 0.72 times the world's present-day energy consumption (see note 5). They are fixed at about 30% of the theoretical heat deposition at the ridges required by rates of sea floor spreading (i.e., heat released by magma constantly cooling at boundary) inferred from radiometric dating of sea floor rocks. See J. E. Lupton, E. T. Baker, and G. J. Massoth, "Variable ^{3}He/Heat Ratios in Submarine Hydrothermal Systems: Evidence for Two Plumes Over the Juan de Fuca Ridge," Nature **337**, 161 (1989); C. A. Stein and S. Stein, "Constraints on Hydrothermal Heat Flux through the Oceanic Lithosphere from Global Heat Flow," J. Geophys. Res. **99** (B2), 3081 (1994); H. Elderfield and A. Schultz, "Mid-Ocean Ridge Hydrothermal Fluxes and the Chemical Composition of the Ocean," Annu. Rev. Earth Planet. Sci. **24**, 191 (1996).

419. No one has yet drilled in the ocean for geothermal heat, presumably because the cost is not justified. However, the land-based technology is fairly well advanced. See J. W. Tester *et al.*, "The Future of Geothermal Energy: Impact of Enhanced Geothermal Systems (EGS) on the United States in the 21st Century," Massachusetts Institute of Technology and Idaho National Laboratory, INL/EXT-06-11746, 1 Jan 06.

420. The most notorious of these earthquake scares occurred in Basel, Switzerland, in 2006. See K. Harmon, "How Does Geothermal Drilling Trigger Earthquakes?," Scientific American, 29 Jun 09; J. Glanz, "Geothermal Drilling Safeguards Imposed," New York Times, 15 Jan 10.

421. There are serious plans in the works to tap the kinetic energy of the Gulf Stream. See A. Ansari, "Is the Ocean Florida's Untapped Energy Source?," CNN, 27 Jul 09. See also J. Copping, "Ocean Currents Can Power the World, Say Scientists," The Telegraph, 29 Nov 08.

422. Estimates of the Gulf Stream velocity vary widely but are never greater than 2 meters per second. Its Pacific analog, the Kuroshio Current, has maximum velocities of about 1 meter per second. See W. E. Johns *et al.*, "Gulf Stream Structure, Transport, and Recirculation Near 68° W," J. Geophys. Res. **100** (C1), 817 (1995); M. Feng, H. Mitsudera, and Y. Yoshikawa, "Structure and Variability of the Kuroshio Current in Tokara Strait," J. Phys. Oceanography **30**, 2257 (2000). If the world's oceans (0.7 of earth's surface, depth 4,000 meters) all had local velocities of 2 m/sec, the total kinetic energy of the ocean would be $1/2 \times 10^3$ kg/m^3 × 0.7 × 4π × $(6.37 \times 10^6$ m$)^2$ × 4,000 m × $(2$ m/sec$)^2 = 2.86 \times 10^{21}$ joules. This is 6.5 times the world's present annual energy use (see note 5).

423. Ocean Thermal Energy Conversion (OTEC) technology is bedeviled by friction losses in its pipes. See K. Galbraith, "Generating Energy from the Deep," New York Times, 29 Apr 09. See also W. H. Avery and C. Wu, *Renewable Energy from the Ocean* (Oxford U. Press, 1994); M. Ravindran, "Harnessing of the Ocean Thermal Energy Resource," in *Oceanology*, H. K. Gupta, ed. (Universities Press, 2005).

424. Even with the very generous assumptions that the tropical ocean surface temperature is 26.5°C and the bottom temperature is 0°C, the ideal Carnot efficiency of Ocean Thermal Energy Conversion engines is only $1 - (273°K/300°K) = 0.09$ (see note 423). The net useful energy content of a cubic meter of water is thus 0.09×10^3 kg/m^3 $\times 4.18 \times 10^3$ joules °K^{-1}kg^{-1} $\times 26.5°K = 1.00 \times 10^7$ joules/m^3. The flow of such water required to power the world is thus 4.4×10^{20} joules/y / (1.00×10^7 joules/m^3) = 4.4×10^{13} m^3/y (see note 5). This is 4.4×10^{13} m^3/y / (6.68×10^{12} m^3/y) = 6.6 times the average flow of the Amazon (see note 49) and 4.4×10^{13} m^3/y $\times 6.29$ bbl/m^3 / (2.0×10^6 bbl/d $\times 365$ d/y) = 3.79×10^5 times the peak flow of the Alaska Pipeline (see note 162).

425. The tides most often identified as potential power sources are the mighty sixteen-meter ones in the Bay of Fundy. However, the cost of the generators has always been problematic, especially given Nova Scotia's ample supplies of conventional hydropower. See "Fundy Tidal Power Demonstration Approved," CBC News, 17 Sep 09. The most aggressive exploitation of tidal power at present is in Britain, and even that is modest—so far. See A. Jha, "First Tidal Power Turbine Gets Plugged In," The Guardian, 17 Jul 08; "Tidal Energy System on Full Power," BBC News, 18 Dec 08; J. Vidal, "Severn Tidal Power Scheme Should Not Go Ahead, Warns Environment Agency," The Guardian, 17 Jul 09; T. Webb, "Islay to Be Entirely Powered by Tides," The Guardian, 25 Aug 09. See also I. Ordóñez, "Everybody into the Ocean," Wall Street Journal, 6 Oct 08; B. Walsh, "Catching the Currents," Time, 15 Jan 09; L. Blumenthal, "Obama Seeks Funding Cuts for Wave, Tide Energy Research," McClatchy Newspapers, 31 May 09.

426. Records of solar eclipses from ancient China and Rome show that earth's rotational period has increased an average of 32 sec/century^2 $\times$ (10^{-2} century/y)^2 $\times$ 2 / (1 d $\times$ 365 d/y) = 17.5×10^{-6} seconds per year. Theory based on modern laser ranging observations of the moon suggests a slightly higher value of 23.0×10^{-6} seconds per year. (The disparity is often attributed to glacial unloading.) The earth's moment of inertia is nominally 2/5 $\times 5.98 \times 10^{24}$ kg $\times$ (6.37×10^6 m)^2 = 9.71×10^{37} kg m^2. The power the earth puts into the tides is thus less than 9.71×10^{37} kg m^2 $\times$ $(2\pi)^2 \times 2.3 \times 10^{-5}$ sec/y / (1 d $\times$ 24 h/d $\times$ 3,600 sec/h)^3 = 1.37×10^{20} joules per year, or about 0.3 times the present world energy budget (see note 5). The most commonly quoted number is about half this value. See L. V. Morrison and F. R. Stephenson, "Historical Values of the Earth's Clock Error ΔT and the Calculation of Eclipses," J. Hist. Astron. **35**, 327 (2004); J. O. Dickey *et al.*, "Lunar Laser Ranging: A Continuing Legacy of the Apollo Program," Science **265**, 482 (1994); W. Munk, "Once Again: Once Again—Tidal Friction," Prog. Oceanogr. **40**, 7 (1997).

427. Air compressed to abyssal pressures has density of 387 atm. / (215 atm.) = 1.8 times that of air in a scuba tank (see note 403).

428. The specific gravity of air forced to the bottom of the sea is 0.028 kg/mole × 9.8 m/sec² × 4,000 m / (8.314 joules °K⁻¹mole⁻¹ × 273°K) = 0.484. The energy (buoyancy plus isothermal compression) stored in a given volume (after compression) of air is thus 3.92×10^7 pascals × (1 − 0.484 + ln (388 atm. / 1 atm.)) = 2.54×10^8 joules/m³ (see note 403). The energy stored in a given volume of solar salt heated by 100°K is 2.62×10^6 joules °K⁻¹m⁻³ × 100°K) = 2.62×10^8 joules/m³ (see note 382).

429. The density of liquid natural gas is about 420 kg/m³.

430. Big liquid natural gas tanks are cylinders about 100 meters in diameter and 100 meters tall. They're made this large to minimize the surface-to-volume ratio and thus to minimize heating. The largest tank built to date holds 2.0×10^5 m³. However, several larger tanks are presently under construction, and even larger ones are on drawing boards, so this record will grow. See Y.-m. Yang *et al.*, "Development of the World's Largest Above-Ground Full Containment LNG Storage Tank," Korea Gas Corporation, 5 Jun 06 (Paper presented at the 23rd World Gas Conference, Amsterdam, 5–9 Jun 06); O. Tsukimori, "Tokyo Gas Says to Build World's Biggest LNG Tank," Reuters, 26 Apr 09. Some useful case study numbers are available at the Lusas Engineering Software website (www.lusas.com).

431. Assuming for simplicity that the tanks in question have diameters and heights of 100 meters, they have volumes of π × (50 m)² × 100 m = 7.85×10^5 m³ and store energy 7.85×10^5 m³ × 2.54×10^8 joules/m³ = 1.99×10^{14} joules (see note 428). The number of such tanks required to power the Los Angeles Metropolitan Area for one day is thus (1.41×10^{10} watts × 3,600 sec/h × 24 h) / (1.99×10^{14} joules) = 6.12 (see note 381). Each tank requires area of $3^{1/2}$ /2 × (100 m)² = 8.66×10^3 m² when closely packed. The total seafloor area required to store the Los Angeles Metropolitan area for one day is thus 6.12 × 8.66 × 10^{-3} km² = 0.053 km².

432. The number of large undersea compressed air tanks required to store the energy of Lake Mead (see note 199) is 7.68×10^{16} joules / (1.99×10^{14} joules) = 386 (see note 431). The amount of ocean floor required for these tanks is 386 × 8.66 × 10^{-3} km² = 3.34 km² or 3.34 km² / (640 km²) = 5.22×10^{-3} times the area of Lake Mead (see note 198). This is a square patch $(3.34 \text{ km}^2)^{1/2}$ = 1.82 kilometers on each side.

433. According to the U.S. Energy Information Agency's "International Energy Annual," the world consumed 1.6×10^{13} kwh × 3.6×10^6 joules/kwh = 5.76×10^{19} joules in 2006. The number of undersea compressed air tanks required to store this energy is 5.76×10^{19} joules / (1.99×10^{14} joules) = 2.89×10^5 (see note 431). The amount of ocean floor required for these tanks is 2.89×10^5 × 8.66 × 10^{-3} km² = 2.50×10^3 km². This is a square patch $(2.50 \times 10^3 \text{ km}^2)^{1/2}$ = 50 km on each side.

434. The number of undersea compressed air tanks required to store the total energy needs of the world for one year is 4.4×10^{20} joules / (1.99×10^{14} joules) = 2.21×10^6 (see notes 5 and 431). The amount of ocean floor is 2.21×10^6 × 8.66 × 10^{-3} km² = 1.91×10^4 km². This is a square patch $(1.91 \times 10^4 \text{ km}^2)^{1/2}$ = 138 km on

each side. The areas of New Hampshire and Israel are 2.42×10^4 km^2 and 2.20×10^4 km^2, respectively.

435. The compressed air tank area required to store the total energy needs of the world for one year is 2.21×10^4 km^2 / $(0.7 \times 4\pi \times (6.37 \times 10^3$ km$)^2) = 6.19 \times 10^{-5}$ times the total ocean floor area (see note 434).

436. Only two actual Compressed Air Energy Storage (CAES) plants actually exist at the moment: Huntorf, Germany (290 megawatts), built in 1978, and McIntosh, Alabama (110 megawatts), built in 1991. Both use salt dome caverns (see note 194). See D. R. Richner *et al.*, "Solution Mining: In Situ Techniques," in *SME Mining Engineering Handbook*, 2nd ed., H. L. Hartman, ed. (Society for Mining, Metallurgy and Exploration, 1992); G. Grazzini and A. Milazzo, "Thermodynamic Analysis of CAES/TES Systems for Renewable Energy Plants," Renewable Energy **33**, 1998 (2008); J. K. Warren, *Evaporites: Sediments, Resources and Hydrocarbons* (Springer, 2006); R. Kushnir, A. Ullmann, and A. Dayan, "Compressed Air Flow within Aquifer Reservoirs of CAES Plants," Transport in Porous Media **81**, 219 (2010).

437. Methane at abyssal pressures and temperatures is still far above its critical point, so its compressive energy is the same as air's (see note 428). Its specific gravity, however, is $16/28 \times 0.484 = 0.277$, so its buoyancy energy is correspondingly greater than air's. To the buoyancy and isothermal compressive energy we also add the heat of combustion. The total energy per unit volume of the abyssal compressed methane is thus $(3.92 \times 10^7$ pascals $\times (1 - 0.277 + \ln (388$ atm / 1 atm$)) + 277$ kg/m$^3 \times 5.5 \times 10^7$ joules/kg$) / (2.54 \times 10^8$ joules/m$^3) = 61.0$ times the energy per unit volume of compressed air.

438. The U.S. Strategic Petroleum Reserve has a total capacity of 727 million barrels (www.spr.doe.gov/dir/dir.html). This volume is equivalent to 7.27×10^8 bbl / $(6.29$ bbl/m$^3 \times 7.85 \times 10^5$ m$^3) = 147.2$ large undersea compressed air tanks (see note 431). This is $147.2 / 386 = 0.38$ times the number of such tanks required to store the energy of Lake Mead (see note 432).

439. According to the U.S. Department of Energy's natural gas website (www.eia.doe.gov/naturalgas), the total U.S. underground storage capacity for natural gas in 2009 was 8.7×10^{12} ft^3 / $(3.28$ ft/m$)^3 = 2.47 \times 10^{11}$ m^3 (uncompressed). If this amount of gas were compressed to a pressure of 388 atmospheres, as appropriate for the abyss, it would occupy the volume of 1 atm $\times 2.47 \times 10^{11}$ m^3 / $(388$ atm $\times 7.85 \times 10^5$ m$^3) = 811$ large undersea compressed air tanks (see note 431). This is $811 / 386 = 2.10$ times the number of such tanks required to store the energy of Lake Mead (see note 432).

440. Each large undersea compressed air tank (see note 431) stores the energy equivalent of 1.99×10^{14} joules / $(6.27 \times 10^{13}$ joules$) = 3.17$ Hiroshima-sized atomic bombs (see note 189).

441. Compressed air escaping from a tank in the abyssal deep and rising to the surface releases $\ln(388) / (1 - 0.484 + \ln(388)) = 0.92$ of its stored energy in cooling seawater and forming ice (see note 428). The ice latent heat of fusion is 3.34×10^5

joules/kg. If all the energy in a catastrophic undersea tank failure available for cooling makes ice, the amount of ice formed is thus $0.92 \times 1.99 \times 10^{14}$ joules / (3.34×10^5 joules/kg) = 5.48×10^8 kg (see note 431). The U.S. Coast Guard defines a "medium sized" iceberg to be 90 meters long and 30 meters tall. (See the FAQ at www.uscg-iip.org.) Assuming that the visible part of such an iceberg is a cylindrically symmetric pyramid and that 90% of the iceberg's ice is underwater, the iceberg has a total volume of $\pi/3 \times (45 \text{ m})^2 \times 30 \text{ m} \times 10 = 6.36 \times 10^5 \text{ m}^3$ and total mass of $6.36 \times 10^5 \text{ m}^3 \times 917 \text{ kg/m}^3 = 5.83 \times 10^8$ kg.

442. Both the amount of gas emitted in the Lake Nyos event and the size of the fountain are estimates. The event unfortunately occurred at night and wasn't witnessed. Lake Nyos is located in Cameroon. It is quite small (area of 1.6 km², about half that of New York's Central Park) but very deep. Geothermal springs at the lake's bottom charge the water with carbon dioxide, which stays there under pressure, like fizz in a soda bottle. On August 21, 1986, for reasons that remain controversial, the waters overturned. Water rising to the top, relieved of its pressure, then belched out a huge volume of CO_2. This gas, being heavy, then flowed down a nearby valley and caused mass death by asphyxiation. The thoroughness and reach of the killing (the most distant death was 26 kilometers away) led to gas volume estimates of about 1 km³. The lake the next morning was brown from the overturning, and there was wave damage 20 m high on one shore. See G. W. Kling *et al.*, "The 1986 Lake Nyos Gas Disaster in Cameroon, West Africa," Science **236**, 169 (1987); S. J. Freeth and R. L. F. Kay, "How Much Water Swept Over the Lake Nyos Dam During the 1986 Gas Disaster?," Bulletin of Vulcanology **53**, 147 (1991); M. Holloway, "Trying to Tame the Roar of Deadly Lakes," New York Times, 27 Feb 01.

443. A comprehensive library of bathymetric maps, data sets, and references is maintained by the U.S. National Geophysical Data Center at www.ngdc.noaa.gov.

444. Energy extraction nominally works by adiabatic expansion to a lower pressure, followed by isobaric warming back to ambient temperature. Assuming that the latter is 4°K above the freezing point of (salt) water, and that the gas specific heat ratio is 7/5 = 1.4, the maximum safe elevation rise is $(1 - (271°\text{K}/275°\text{K})^{7/2}) \times 4{,}000 \text{ m} = 200$ m.

445. The number of 4°K expansion stages required to reach the continental shelf (150 meters) is $\ln(150 \text{ m} / 4{,}000 \text{ m}) / \ln(1 - 200 \text{ m}/4{,}000 \text{ m}) = 64$ (see note 444). This estimate is excessive, however, because a chain of 40 stages reaches a depth of 500 meters, the ocean's thermocline, where the temperature, and thus heat available for expansion, begins to rise. Assuming that the power needs of Greater Los Angeles are spread out over fifty stages, the mass flow of air per stage is $2/5 \times 1.41 \times 10^{10}$ watts $\times$ 0.028 kg/mole / (50×8.314 joules °$\text{K}^{-1}\text{mole}^{-1} \times 4°\text{K}$) = 9.50×10^4 kg/sec. The warming water flow per stage is 1.41×10^{10} watts $\times$ 917 kg/m³ / ($50 \times 4.18 \times 10^6$ joules °$\text{K}^{-1}\text{m}^{-3} \times 4°\text{K}$) = 1.55×10^4 kg/sec. The average mass flow through Hoover Dam is 390 m³/sec $\times 10^3$ kg/m³ = 3.90×10^5 kg/sec (see note 164). According to the

Los Angeles Department of Water and Power (www.ladwp.com), the total capacity of the Los Angeles Aqueduct from the Owens Valley (the original 1913 pipeline plus the second 1970 pipeline) is 10^3 kg/m^3 × (485 ft^3/sec + 290 ft^3/sec) / (3.28 ft/m)^3 = 2.20 × 10^4 kg/sec.

446. The required size of an undersea heat exchanger is determined mainly by the currents available to carry away the heat excess or deficit. Abyssal current drift velocities vary depending on location but are typically between 1 and 6 cm/sec. See B. J. Korgen, G. Bodvarsson, and L. D. Kulm, "Current Speeds Near the Ocean Floor West of Oregon," Deep Sea Research and Oceanographic Abstracts **17**, 353 (1970); F. Camden-Smith *et al.*, "A Preliminary Report on Long-Term Bottom-Current Measurements and Sediment Transport/Erosion in the Agulhas Passage, Southwest Indian Ocean," Marine Geology **39**, M81 (1981); A. Camerlenghi *et al.*, "Ten-Month Observation of the Bottom Current Regime Across a Sediment Drift of the Pacific Margin of the Antarctic Peninsula," Antarctic Science **9**, 426 (1997). This excludes abyssal storms, events during which currents can exceed 51 cm/sec (1 knot). See R. E. Schmid, "Scientists Find Seafloor Storms in Ocean," ABC News, 2 Nov 00. Assuming a conservative value of 2 cm/sec for this current, the heat exchanger area required for a single stage of a storage chain capable of powering the Los Angeles Metropolitan Area is 1.55 × 10^4 kg/sec / (917 kg/m^3 × 0.02 m/sec) = 845 m^2, or a square 29 meters on each side.

447. Assuming that the temperature jump for both extraction and storage is 4°K, the ideal thermodynamic efficiency for a round-trip storage cycle is $(1 - 4°K/273°K)^2 = 0.97$.

448. Taking the temperature of very warm surface water to be about 25.5°C (80°F), the initial safe elevation rise would be $(1 - (271°K/300°K)^{7/2})$ × 4,000 m = 1,200 m (see note 444). The number of such stages required to reach the continental shelf (150 meters) is ln (150 m / 4,000 m) / ln (1 − 1,200 m / 4,000 m) = 9.2. The total flow of such water required to power the Los Angeles Metropolitan Area is 1.41 × 10^{10} watts × 10^3 kg/m^3 / (4.18 × 10^6 joules °K^{-1}m^{-3} × 25 °K) = 1.35 × 10^5 kg/sec.

449. It's less, actually. The boiling temperature of water at one atmosphere is 100°C. The heat required to expand compressed air in an abyssal tank (388 atm.) can be stored in an identical tank filled with water at temperature 2.54 × 10^8 joules/m^3 / (4.18 × 10^6 joules °C^{-1}m^{-3}) = 61°C (see note 428).

450. G. Aloisi, M. Bianca Cita, and D. Castradori, "Sediment Injection in the Pit of the Urania Anoxic Brine Lake (Eastern Mediterranean)," Rend. Fis. Acc. Lincei **17**, 243 (2006); W. Booth, "Oceanography: Offshore River of Brine," Washington Post, 14 May 90; R. F. Shokes *et al.*, "Anoxic, Hypersaline Basin in the Northern Gulf of Mexico," Science **196**, 1443 (1977); D. A. Ross and J. M Hunt, "Third Brine Pool in the Red Sea," Nature **213**, 687 (1967). A video of waves created on one of these lakes by the submersible *Alvin* and the lapping of these waves on the "beach" is linked at H. Roberts, "The Brine Lake," NOAA Ocean Explorer, 31 May 06.

451. The ratio of this lake's area to Lake Mead's is 123 km^2 / (640 km^2) = 0.19 (see note 198). The ratio of its volume to Lake Mead's is 13.3 km^3 / (⅓ × 0.221 km × 640 km^2) = 0.28 (see note 199). It lies at a depth of 2,251 meters. See R. S. Pilcher and R. D. Blumstein, "Brine Volume and Salt Dissolution Rates in Orca Basin, Northeast Gulf of Mexico," Am. Assn. Petrol. Geol. Bull. **91**, 823 (2007).

452. Assuming that both saturated brine and seawater compress the same way, the density difference between them will remain roughly the surface value of 1,203 kg/m^3 − 1,025 kg/m^3 = 178 kg/m^3. The energy storage capacity of a salt lake the size of Lake Mead on the continental shelf is thus (4,000 m/221 m) × (178 kg/m^3 / 1,000 kg/m^3) = 3.22 times the energy storage capacity of Lake Mead (see note 199).

453. Modern deep-drilling muds are made with $CsCO_2H$ brines, which have specific gravities up to 2.3. See W. Benton and J. Turner, "Cesium Formate Fluid Succeeds in North Sea HPHT Field Trials," Drilling Contractor, May/June 2000. Saturated CsCl brine only has a specific gravity of 1.9 at room temperature, but it's much cheaper. The saturation density of CsCl brine at abyssal temperature is 1800 kg/m^3. Were CsCl brine substituted for NaCl brine in an undersea hydro-pump lake the size of Lake Mead, its energy content would thus be increased to (see note 452) 4.30 × (1,800 kg/m^3 − 1,025 kg/m^3) / (178 kg/m^3) = 18.7 Lake Mead energies. At 4°C the solubility of CsCl is about 1.6 kilograms per 1.0 kilograms of water. The mass of CsCl in such a lake would thus be (see note 199) 1/3 × 6.4 × 10^8 m^2 × 221 m × 1,800 kg/m^3 × 1.6 / (1.6 + 1) = 5.22 × 10^{13} kg, and its cost would be 5.22 × 10^{13} kg × \$157 kg^{-1} = \$8.20 × 10^{15}.

454. The number of large underwater brine lakes required to store the world's electricity for one year is (64 meads/month × 12 months) / 4.30 = 179 (see notes 199 and 450).

455. False predictions about when oil will exhaust are legend—as are false predictions that it will never exhaust. See M. Maier, "Ending the End of Oil," CNN Money, 1 Dec 05; C. Krauss, "New Way to Tap Gas May Expand Global Supplies," New York Times, 9 Oct 09; A. Davis, "In a Cheap-Gas World, a Profit Patch," Wall Street Journal, 10 Sep 09.

456. T. Doggett, "US Sees Non-OPEC Oil Output Growth Ending in 2011," Reuters, 12 Jan 10. Good information on how much oil has come out of the North Sea and Alaska's North Slope is hard to come by, but both fields probably had about 80 billion bbls originally and are now more than two-thirds spent (see note 405). See C. Baltimore, "Alaska North Slope May Hold 36 Bln Bbl Oil—US DOE," Reuters, 29 Jan 08; A. Kwiatkowski, "BP Says North Sea Oil, Gas Production Will Drop 9%," Bloomberg, 21 Jul 09.

457. According to the BP Statistical Review of World Energy 2008, the Middle East, which has been very thoroughly surveyed, has 103 billion tonnes (755 billion bbls) of crude oil reserves. "Untapped" reserves there include the Majnoon, Halfaya, East Baghdad, and West Qurna fields in Iraq (45 billion bbls); the Khurais, Abu Jifan, and Mazalij fields in Saudi Arabia (27 billion bbls); and the Ferdows,

Mound, and Zageh fields in Iran (38 billion bbls). Some of these "untapped" fields have recently been tapped. See J. Chen, "The List: The World's Largest Untapped Oil Fields," Foreign Policy, 1 Dec 08; "Giant Saudi Oil Field May Pay at the Pump," CBS News, 29 Jun 08; V. Walt, "Pump It Up: The Development of Iraq's Oil Reserves," Time, 7 Dec 09; "Iraq Oil Field Goes to Royal Dutch Shell and Petronas," New York Times, 11 Dec 09. The recent strikes in Brazil's offshore Carioca, Tupi, Iara, Guara, and Campos fields probably amount to 50 billion bbls (see note 405). The U.S. Geological Survey estimates the Arctic Ocean to contain 90 billion bbls. The Gulf of Mexico probably contains 20 billion bbls more. See E. Douglass, "Chevron Reports Major Oil Field Find," Los Angeles Times, 6 Sep 06.

458. "An Estimate of Recoverable Heavy Oil Resources of the Orinoco Oil Belt, Venezuela," U.S. Geological Survey, Fact Sheet 2009-3029, 1 Oct 09. This report brackets the recoverable Venezuelan oil sands reserves between 3.8×10^{11} bbl and 6.5×10^{12} bbl, with a median estimate of 5.1×10^{11} bbl. That is comparable to all the world's other proved reserves combined. According to the BP Statistical Review of World Energy 2008, those reserves at the end of 2007 were 1.39×10^{12} bbl. Both BP and the CIA World Factbook say that Middle East reserves at the end of 2007 were 7.55×10^{11} bbl. The Canadian oil sands reserves are also comparable to reserves in the Middle East, but how we count them is unclear because it depends critically on the oil market price. See "Alberta's Energy Reserves 2007 and Supply/Demand Outlook 2008–2017," Energy Resources Conservation Board, ST98-2008, 1 Jun 08.

459. BP Statistical Review of World Energy estimates the world proved oil reserves at the end of 2009 at 1.33 trillion barrels and the world oil consumption rate at 84.1 million barrels per day. This gives an oil exhaustion time of 1.33×10^{12} bbl / (8.41×10^7 bbl/d $\times$ 365 d/y) = 43.3 years. The exhaustion time reckoned with 2008 numbers (before the Macondo spill) is 39.5 years. The difference is primarily due to the big undersea strikes in Brazil (see note 457).

460. Increasing the BP reserve estimates by the most optimistic USGS Venezuelan oil sands estimates (see note 458) gives a time to exhaust world oil reserves of (1.33 + 0.65) $\times 10^{12}$ bbl / (8.41×10^7 bbl/d $\times$ 365 d/y) = 75.3 years (see note 459).

461. The BP Statistical Review of World Energy estimates the Middle East proved oil reserves at the end of 2009 at 0.754 trillion barrels and the Middle East production rate at 24.4 million barrels per day (see note 5). This gives an exhaustion time of 7.54×10^{11} bbl / (2.44×10^7 bbl/d $\times$ 365 d/y) = 84.7 years.

CHAPTER 11: A WINTER'S EVE

462. The most dire predictions of global warming over the next two centuries, about 10°C on average, would presumably be noticeable in the U.S. Midwest. See J. Eilperin, "New Analysis Brings Dire Forecast of 6.3 Degree Temperature Increase," Washington Post, 25 Sep 09. However, Murphy's Law guarantees that the amount of warming wouldn't be very great in places we want to be warmer. More

seriously, the dire predictions in this article come from something called the Sustainability Institute (www.sustainer.org), not the United Nations Environment Programme (UNEP). The UNEP report cited in the article makes no predictions about future global temperatures. This is not surprising, because reliably predicting the weather a century from now is technically impossible at the moment. See C. P. McMullen and J. Jabbour, eds., "Climate Change Science Compendium 2009," United Nations Environment Programme, 1 Sep 09.

463. This assumes that one nuclear war is considerably smaller than 40,000 megatons (see note 196).

464. Manufacturers claim that undersea cables can now carry up to 2 gigawatts, but a more reasonable number for long-haul applications is 1 gigawatt (see note 170). According to the New York Independent System Operator (NYISO), New York City's average power consumption is 6.28 gigawatts (see note 154). The number of undersea cables required to power New York City is thus about 6. According to the U.S. Energy Information Administration's "International Energy Annual," Europe's electricity consumption in 2006 was 3,297 billion kilowatt hours. The number of undersea cables required to power Europe is thus 3.30×10^{15} wh/y / $(24 \text{ h/d} \times 365 \text{ d/y} \times 10^{9} \text{ w/cable}) = 377$ cables.

465. The freezing energy of Hudson Bay is comparable to the energy used by civilization and exceeds the latter only if one includes James Bay, Foxe Channel, Foxe Basin, Hudson Strait, and Ungava Bay, thus bringing the total area up to 1.24 million square kilometers. The ice of Hudson Bay freezes and melts completely every year and has an average January thickness of 1.6 m. See W. A. Gough, A. S. Gagnon, and H. P. Lau, "Interannual Variability of Hudson Bay Ice Thickness," Polar Geography **28**, 222 (2004). The energy involved is (see note 441) $1.24 \times 10^{12} \text{ m}^2 \times 1.6 \text{ m} \times 10^{3} \text{ kg/m}^3 \times 3.34 \times 10^{4}$ joules/kg $= 6.62 \times 10^{20}$ joules. The total annual energy budget of civilization is presently 4.4×10^{20} joules (see note 5).

466. The area of Canada is about 9 million square kilometers (see note 46). The area of the arctic ocean is about 14 million square kilometers.

467. Water's latent heat of vaporization is 2.26×10^{6} joules/kg. Assuming that 1 meter of rain falls, on average, on the surface of the earth in a year, the total amount of heat absorbed and released per year in making rain is (see note 7) $4\pi \times (6.37 \times 10^{6} \text{ m})^2 \times 1 \text{ m} \times 10^{3} \text{ kg/m}^3 \times 2.26 \times 10^{6}$ joules/kg $= 1.15 \times 10^{24}$ joules. This is 2,614 times the present energy budget of civilization (see note 5). The latent heat of freezing is much smaller, so the difference between rain and snow is unimportant.

468. This is approximate, and low by some estimates (see note 13). According to the BP Statistical Review of World Energy 2008, the earth had 0.193 trillion tonnes of proved oil reserves and 0.847 trillion tonnes of proved coal reserves in 2007. The BP numbers include neither Venezuelan oil sands (see note 458) nor U.S. oil shale reserves (see note 65).

469. According to the BP Statistical Review of World Energy 2008, the proved oil reserves of the world at the end of 2009 were 1.33 trillion bbls. If 85 million bbls of Mississippi River water flow past New Orleans in 13.2 minutes, then 1.33 trillion bbls

flows past in $(1.33 \times 10^{12}$ bbls $/ 8.5 \times 10^7$ bbls$) \times 13.2$ min $/ (60$ min/h $\times 24$ h/d $\times 365$ d/y$) = 0.39$ year (see note 5). However, this number is probably about a factor of 2 too small because of omission from the report of the Venezuelan oil sands (see note 458).

470. A. Frebel *et al.*, "Discovery of HE 1523-0901, a Strongly r-Process Enhanced Metal-Poor Star with Detected Uranium," Astrophys. J. **600**, L117 (2007).

INDEX

Telecommunications, undersea,
     106–107
Termites, 77
Terrorism, dirty bombs and, 57
Texas
     California energy crisis and Texas
          energy companies, 22–23
     deregulation of power rates and, 28
Thermal energy, 51
Thermal storage, 94–97
     energy capacity of, 96–97
Thermohaline circulation, 108
Thorium, 61
Three Gorges Dam, 47
Three Mile Island, 57, 59
Tides, as energy source, 108
Trash, 79–89, 116, 119
     ash pollution and incineration of,
          85–88
     carbon in, 80–82, 83–84
     disposal undersea, 107
     electric car batteries and, 88–89
     energy yield from incineration of,
          80–81
     environmental trouble from
          incineration of, 83–87
     metals in, 85–86, 87–89
     poisons released by incomplete
          incineration of, 84–85
     as price regulator, 80, 83, 89
     synthesis gas from, 81–82, 83, 84
Trash gasification startups, 82
Triassic, 14
Trichoderma reesei, 75
Turbine technology, 93
Twain, Mark, 76, 116
TXU, 28

U-235, 58
U-238, 58–59, 61, 63
Undersea energy storage, 108–113
Undersea power cables, 48

Undersea telecommunications
     companies, 106–107
U.S. Energy Information Agency, 113
U.S. Federal Energy Regulatory
     Commission Order No. 888, 22,
     23
U.S. Strategic Petroleum Reserve, 109
Uranium
     extraction of, 59–60
     isotopes of, 58–59, 61, 63
     role in future energy scenario,
          118–119
     supply of, 57–58
Uranium atom, 33

Venezuelan oil sands, 113
Victoria Falls, 47
Virginia, regulation of power rates in, 28

Wall Street Journal (newspaper), 95
Water, electrolyzing, 94
Waterfall, power of, 47
Water hardness, 86
Water pumping, 51–52, 53–54
Water quality, leaching metals and, 86,
     87
Weald, age of, 13–14
Weather patterns, climate change and, 8
"Wheel-out" scheduling scheme, 27
White cliffs of Dover, 14
Wind energy, 93–94, 98
     energy future and, 3
     energy storage and, 53
     energy yield from, 93, 98
     marginal demand and, 49
Withholding power, California energy
     crisis and, 26
Wood chips, 75, 81

Zinc
     in batteries, 88
     health and, 86